OUT IN THE FIELD:

Discovering a Career in Field Biology

By Tim E. Hovey

For more information visit:

WWW.TIMHOVEYBOOKS.COM

Library of Congress Cataloging-in-Publication Data

Tim E. Hovey

Out in the Field/Tim E. Hovey

p. cm.

ISBN: 978-1475182422

10 9 8 7 6 5 4 3 2 1

To my dad:

I miss you every day. Although you weren't alongside me for many of these adventures, you're with me on all of them now.

He didn't tell me how to live; he lived, and let me watch him do it.

~Clarence Budington Kelland

Table of Contents

PREFACE

If someone had sat me down as a child and explained to me exactly what a field biologist did, I'm pretty sure I would've thought it was the greatest job in the world. If they continued to explain the training would involve intense classes in physics, chemistry and calculus, I'm sure my young brow would have been furrowed; as I would have had no idea what those things were. But none of that happened. There was no example in my family to follow, no childhood interaction with a forest ranger or a warden that sparked the flame of interest, nor was there an understanding of what a career working in the wild could mean. There was only that clear-as-day determination to be outside.

For as long as I can remember I have been drawn to the outdoors. All my fondest childhood memories are of camping near coastal lakes, exploring sycamore-choked creeks, and throwing slimy things at my brother, Steven. And early exposure to this lifestyle and the ability to pull on the tether of adventure would not have been possible without my parents. As a kid, regular family camping trips were as exciting as Christmas for me and my brother. My mom and dad would load us into the cherry-red, 1970 Volkswagen bus and we would head to wherever the road ended and the land was flat enough to park.

We explored the countless campgrounds along the coastal range of Santa Barbara, Santa Ynez, and San Luis Obispo counties; usually punctuating those trips with a family gathering in central California in the fall. It was a gift my parents gave me and my brother that was beyond wealth. It was during this early time the seed of adventure was planted.

Most of our outdoor adventures focused around water, be it a freshwater lake or the sea. Even now, I wonder if this introduction to the aquatic world channeled my interests towards fisheries, or if I was simply born to do what I do. Regardless, it was during one of these early outings that I realized a different world existed beneath the water's surface.

Sitting on the muddy bank of a lake, I caught my first fish at the age of five. I clearly remember dragging the small bluegill ashore with the thick, green drop line; the fish flopping behind the red and white bobber covered with dirt. I knelt aside it in the mud and looked at the little pan fish. I remember feeling an almost unimaginable excitement. As I reached down and touched its smooth fins, something clicked. Thinking back on it now, it was that sunny morning on the bank that changed my life.

Over time my interest in the outdoors and fishing went well beyond just a casual enjoyment. I read everything I could on fishing and fish, and every camping trip into the outdoors seemed to teach me something new. At the time, however, I was too young to understand this interest was anything more than a hobby. I soon found it was gradually becoming much more than just an interest.

As many do when adulthood neared, I simply followed in my father's footsteps when it came time to choose a career. He was a mechanical engineer, and while it was painfully clear he didn't enjoy what he did for a living, I figured that was just part of being an adult. After high school I worked for several engineering firms as a drafter in the Santa Barbara area. I gradually learned the subtle intricacies of an electronic schematic and my artist's eye, passed to me by my father, made drawing any mechanical part a breeze. I began taking night classes at a community college in my early twenties. The plan was to eventually get a degree in engineering. Yet, always in the back of my head, was that nagging feeling I should be doing something else.

During a career fair at the college, I found myself reading several biology handouts distributed by one of the many four-year colleges that attended in hopes of convincing young students to transfer to their institute. After a few moments of reading, I found what I was looking for. The classes required for a transfer to the marine biology program at California State University at Northridge, looked strangely familiar. With the exception of one class, all the classes I had taken so far would easily slide over and count for the general education requirements. That left two years of core classes to complete and my drawn-out college career would be over. I can clearly remember smiling as I glanced over the leaflet.

I had unknowingly prepared for the career I was programmed to have practically since birth, while taking classes for a degree towards a career I couldn't stand. Since I was already well-established in the engineering field with a good job, turning my back on all that and becoming a full-time student was going to take some thought.

The push to make the switch came a few short weeks later while working with Al, an older engineer with the company I was currently working for. I was assigned to assist him during a small project and during that time, I came to two very powerful conclusions: I had no desire to be an engineer and no desire to end up like Al.

In the summer of 1990, I was accepted to the marine biology program at the university. Before my first year of college was complete, I had secured a few volunteer spots for summer research voyages and was fortunate enough to meet Dr. Larry Allen, head of the fisheries program. After a first impression that clearly demonstrated I was bothering him, he quickly took me under his wing once he learned of my direction and love of fisheries science. At the time of our meeting, I had no idea how important Dr. Allen's friendship and guidance would be in guiding my career.

I can count on one hand the number of influential people in my life and usually have a few fingers left over. First and foremost is my father. His friendship, patience, and ability to guide and not steer sets him at the top of the list, safe from ever being dethroned. A close second is my friend Larry.

When that first summer research trip rolled around, I was determined to learn whatever I could on every voyage. I learned the way of the sea and was quickly schooled on what was expected of the volunteers on the confines of a vessel. If you slept and ate on the boat, you also worked. There was no room for slackers. As the different crews came and went, those that weren't cut out for the field tasks were easy to spot. They usually didn't last very long and Larry, Danny and Jimmy (the principle scientist, chief engineer, and vessel captain respectively) never even bothered learning their names. I can proudly say that if I walked into their offices today, they would know who I am.

With experience came more responsibility. As the trips piled up, I found that more important tasks were being assigned to me during research voyages. I felt privileged that I had wedged myself into the training of a fisheries biologist and was slowly moving up the ladder. To increase my marketability on and around the research vessel, I became scuba and small craft certified. Soon afterwards, I was running the small work boats that I had earlier only been allowed to ride on.

I immersed myself in the training, feeling almost as though I was living someone else's life. The early boat trips and the people I was meeting made me realize I was finally on the right path. Unfortunately, even the right path may at times be dangerous. While in college, I had to deal with the death of a good friend at the hands of the sea. I had met my friend, Steve, on the deck of that same research vessel on which I had trained. He was an experienced free diver and more than generous with his knowledge of the sport when I decided I wanted to start free diving. Steve taught me to never turn my back on the ocean, and now as I struggled with his passing, it seems that he had briefly ignored his own advice.

On graduation day, only an hour after receiving the diploma that I had chased for more than seven years, I was on a plane to Austin, Texas to attend a fisheries conference. As I mingled with the scientists at the meeting hall, I understood that even though I had received my college degree only hours before, the true training was just beginning.

During that next summer, I accepted a graduate position in Larry's program and prepared for the rigors of an advanced degree. The three remaining graduate positions were scooped up by my three closest friends. Now, focused research trips and sample collecting would be for our own projects and we would be the leads. We would run the boats, pick the volunteers, assign the tasks – we would run the show.

The next two years were filled with countless trips for the scientific benefit of our projects and adventures beyond anything we could imagine. We followed the road south into Baja California and camped wherever we were when the sun went down. We freely lived off the land, catching and spearing whatever we needed from the sea. And always, we collected the data. At the end of my graduate project, I had caught and worked up almost two thousand spotted sand bass for the sake of science.

During my tenor at Northridge, I learned to run boats in rough water and dive for my own food. I got to chum the coast for sharks and have been electrocuted 25 feet below the surface by a fish. I have deployed all manner of gear from small skiffs to collect fish for science, and on some occasions, I have used those same small skiffs to run over that gear. I have unknowingly helped cook up and consume a world record fish and I have vomited underwater. I drove the dusty roads of Baja California and got to see some amazing things. As my scholastic training came to a close, I could only imagine how the next chapter would unfold. I actually believed that the adventurous part of my new career was coming to an end. I was soon to find out that this could not have been further from the truth.

When I graduated with my Master's degree, I knew that I was at the end of that gray area of life when an adult could still focus on school and defer thoughts of a career. As I descended the stage with my second diploma in hand, I can remember feeling very sad and extremely excited at the same time. I knew it was time to enter the field.

My time at Northridge, and more importantly my training under Dr. Larry Allen, changed my life forever. As a young adult, I had wandered in an unfocused fog for years, unclear of my path or what might lie ahead. I knew I had an interest, but I did not possess the tools to sharpen my skills and capture a career. My training under Larry gathered all those loose ends and tied them neatly into a direction and, more precisely, a way of life. There in the hallways of the science buildings, I believe it was a Thursday, I became a field scientist.

As my life moves forward, I feel like I learn something new on a daily basis. I treat every bit of information like a gift and I store it away for future use. I was blessed with an unforgiving memory and once I commit any information to it, it is there for good. I have taken my bag of knowledge from Larry and Northridge and I have built a career in the outdoors. I am confident it is where I belong. Despite scars from the journey, I've emerged from a pathway lit by fond memories and gratifying accomplishment. Despite there having been dangerous dips and confounding, frustrating periods, I see only the good times and the adventure. Good and bad, this is how I got here, and I wouldn't have changed a thing.

THE BEGINNING

The gun-metal gray research vessel swayed gently on the anchor at the edge of a large kelp bed just off the coast. She sat like a tethered beast waiting to be set free. The morning fog didn't stand a chance against the clear skies and the beating sun. Out towards the horizon, the ocean looked like gray oil as it heaved and fell with the slight swell. For a few seconds, just a little longer than normal, it was completely quiet. Then the sounds of ship and sea began to fill the emptiness.

During the summer of 1991, I found myself standing on the back deck of a research vessel just off the coast of California. That was my second trip out as a volunteer and I can remember feeling excited about the new direction my life had taken. A year earlier, to the day, I had been employed as a computer-aided designer for a valve company. I would spend my days in a cramped, windowless computer lab staring at a screen. Frustrated with the direction, I turned my back on that lucrative job to study fish.

My path to the deck of the *Yellowfin* and the start of a new career in the field sciences was a jagged one. After high school I had no ambitions to attend college. During my senior year, I'd taken a job as a mechanical and electrical drafter for a company that manufactured printed circuit boards. The money was good for a guy my age and I felt no need to continue my education. Over the next seven years I learned what I could. Gradually, over that time, I realized I didn't like what I was doing. I began taking night classes at the local community college. I wasn't really excited about my career direction in general, but it was at least a direction. I kept to the path, but with little enthusiasm. That is, until I started working with an engineer named Al.

Al was a lead engineer for a company I had been working for. He was an older gentleman, only one year away from retirement. I had been assigned to assist him in the design and production of some part I can't even remember. Early in the task, he called and asked me to meet him at his cubicle to discuss the project. This was the first time I had ever worked with Al and when I rounded the corner to his cluttered, three-wall office for the first time, my soul went cold. Al was sitting in the only chair in the small space because there was no room to stand. The desk was barely visible under scads of documents and engineering magazines. There was one narrow passage to his desk, most of the area stacked with folders, files and clutter. I looked at Al, who hadn't seen me approach the door to his cave. I froze. The thought of ending up like that frightened me. After our meeting, I can remember walking the halls back to my office thinking it was time for a change.

While the decision to make the career change was sudden, the direction of that change was not. For the past few years, I had been investigating a new career path far outside the confining walls of an office building. For as long as I can remember the outdoors, and more precisely the sea, had maintained an almost hypnotic hold over me. I would read anything I could get my hands on regarding the outdoors, the ocean and science. I'd read about the people who studied the sea and all the animals that lived there. How these field scientists would travel the world's oceans and conduct research in onboard labs lined with microscopes, aquariums, and portholes. Even after reading all I could, I only had a slight understanding of what a field biologist actually did. And I had absolutely no idea of how to get there from where I was.

A week after my visit with Al, I took all the credits I had accumulated during my night classes and applied to the marine biology program at California State University at Northridge. After a few months of jumping through hoops, I finally received my acceptance letter. Only two months after my pivotal meeting with Al, I found myself scheduled to attend the fall semester. Before my first full year of school came to a close, I was able to secure a few volunteer spots for the upcoming summer research trips aboard the research vessel, *Yellowfin*.

The 85-foot *Yellowfin* was a perfect work platform for the coastal surveys and would serve as the mother ship for the week long collecting trips. The vessel had a huge open back deck and a wet lab station that stretched along the port side. Two 18-foot work boats floated at the stern, tied off to a large A-frame mounted to the rear of the ship.

During that second summer trip we were collecting larval fish using beam trawl nets deployed from the smaller work boats. The nets had a steel, six foot-wide rectangular opening at the front. A towing bridle was tied to the frame and attached to the back of the skiff during sampling. Behind the frame, a dozen feet of catch net trailed the trawl capturing anything that flowed through the frame mouth. Our job was to deploy the trawls at predetermined depths and tow them for fifteen minutes in the hope of collecting larval fish. After we completed a survey run, we'd head back to the *Yellowfin* and sort out the samples.

Being new to all this, I was still very much in awe of most of the things I was seeing and experiencing. This new enthusiasm was likely the reason I got asked back out by the project's lead investigator, Dr. Larry Allen. The work was hard, and at times tedious, and definitely far from what I was used to. That didn't matter to me. I was being trained for a career in something I had a true interest in, and no matter what they threw at me, I was eager and ready to catch everything.

After we completed the trawls, we collected the sample buckets and gathered around the wet lab to sort them. The larger fish were identified, measured and tossed back overboard alive. We made notes of any invertebrates that were caught and recorded the waypoint and water temperature of every sample. Any larval fish were separated and carefully scrutinized for identification. That was the first time I realized that most tiny larval fish look absolutely nothing like their adult parents.

Tom, one of the lead graduate students on the trip, carefully examined several of the larval specimens we had collected. After searching through each of the six trawl samples, he carefully netted out a single larval specimen of the target species. He transferred the tiny fish into a sample vial, secured the cap and handed it to me. The small white sea bass drifted aimlessly in the tiny vial. The fish had huge eyes and was about half the size of a grain of rice. It was covered with cinnamon colored specs and its huge pectoral fins flapped feverishly in its tiny glass home.

As Tom cleaned up the area, he handed me a heavy plastic writing sheet and a pair of scissors. "Make sure you cut a label and slip it in the sample vial," he said. To keep all the records straight, each specimen vial included a small slip of waterproof paper with all the sample specifics written in pencil. It seemed beyond simple.

I wrote down all the information and cut the stiff paper into a small, inch-and-half square. I carefully placed the vial in one of the racks and loosened the cap. I rolled the small slip of paper into a tiny tube and grabbed the glass vial. I held it up and watched the tiny fish float and bob in its enclosure. There was not a lot of room in the tiny vial, but I knew the specimen had to be labeled. I slid the rolled-up plastic paper into the vial and tapped at it until it slid inside. As soon as the paper cleared the confines of the vial's throat, it unwound and expanded, smashing the tiny fish against the inside of the glass and smearing it like a bug on a windshield. Bits of fin and tail floated to the bottom of the glass casket and any resemblance of a fish was completely gone. Panic flashed in me.

I had no words. Tom had entrusted me with the prized specimen and I had reduced it to a pile of very small bones and scales at the bottom of its container. I stood there completely numb. Tom walked up and asked for the vial. I slowly handed it to him. I watched him hold it up and look for the fish. "Is he still in here?" he asked, shifting the vial in his fingers. I was trying to describe what had happened, when he figured it out for himself. "Ah, you squished him," he added. "No worries," he said, as he lightly patted my shoulder.

The tiny specimens we were collecting were for a genetic study on white sea bass. Each labeled sample would eventually be transported to a genetics lab, and macerated for analysis. As long as they were positively identified as larval white sea bass when labeled, it really didn't matter what shape they arrived to the lab in.

Once the day's work was completed, we were all allowed to enjoy the limited activities in the confines of the ship. Scuba tanks and wetsuits were neatly arranged on the lower deck, placed well outside the work area of the large A-frame that sat like a huge stinger at the back of the vessel. The huge six-foot-wide winch that fed cable through the A-frame sat in the center of the mid deck, greasy and mean. During operation, no one was allowed near the huge spool of thick cable. However, since the beast was now sleeping, all aboard found it a convenient place to lean the dozens of fishing rods in various sizes and colors.

I can remember driving to the dock for my very first research trip for the university and seeing the mass of fishing tackle stacked near the center of the vessel. A smile stretched across my face as I approached the huge gray ship that would be my home for the next seven days. That warm feeling was quickly stomped out when the ships engineer, Danny Warren, saw me approach with my fishing rods. "Oh perfect, more fishing gear", he said, squinting against the smoke from his own cigarette. "It's not like we don't have enough". He lifted a hatch and gruffly disappeared below deck to attend to some unseen duty. My smile was gone, and as I boarded and walked around the open hatch, I wondered how much available air was in the little chamber below when the hatch door was closed and locked from the outside.

With this sampling station complete, I helped clean up the work area, wash down the fish buckets and stow the survey gear. Tom collected all the data sheets and the vial of fish parts and stowed them in one of the gear boxes. A few of the group dispersed to other relaxing activities around the vessel. I grabbed my fishing gear and headed towards the back of the boat. As I dug around for tackle, Lisa, another graduate student, came down to the back deck and started pulling dive gear out of her bag. A few minutes later Jimmy, the captain, came down to the back of the boat with his dive gear. He dropped his bag near the A-frame and looked over to Lisa. "Well, are you ready?" he asked. The State of California had a requirement that in order for students to regularly dive off the research ship, they needed to go through a check out dive with the captain first. From what I gathered of the conversation, the requirements of the check out dive were minimal.

I watched as both divers squeezed into their gear and load themselves down with dive belts, knives, abalone irons and gauges. They assisted each other with the donning of their heavy tanks and finished clicking buckles and adjusting straps. They leaned over carefully grabbing their fins and made their way over to the swim step. They looked uncomfortable and completely overloaded. As Lisa bent down to get into the water, the lanyard on her brand new abalone iron caught on something, snapped and dropped overboard. The bright yellow handle was visible below the surface for a few seconds before it disappeared into the depths. Lisa looked back towards Jimmy. He just shook his head. "Sorry," he said, "we're in a hundred and four feet of water and I'm not supposed to take you passed sixty." She pounded her fist on the ship in anger. The new iron was gone.

The two divers eased into the water and floated on the surface for a few seconds before slowly submerging. I watched as they swam away from the ship and dropped from sight about 20 feet under water. I followed their air bubbles as they popped to the surface through the kelp. Shortly after that, any evidence of the divers was gone. Standing there, now alone, I remember thinking that eventually I would have to muster the courage to get down there myself.

I spent the next 30 minutes fishing at the back of the boat without a bite. The inactivity and the slow rocking of the ship had me daydreaming. I felt myself becoming envious of the divers in the water. At that time, I had absolutely no experience with swimming in the ocean. I do remember that the idea of floating around in the sea scared the hell out of me, but I also understood that I would eventually have to conquer that fear.

A flash several feet underwater caught my attention. A small school of pacific mackerel moved in unison at the edge of the kelp, their silver and green sides flashing like metal as they swam. I looked into the calm water behind the ship. The sunlight pierced the depths and disappeared in subtle, white shards 20 feet below. I knew that eventually I would need to get down there.

A splash near the kelp startled me. I turned just in time to see something just beneath the surface disturb the water. I walked over to the starboard side of the ship for a better view. I scanned the area where I had seen the splash. Inside the kelp bed a dark black fin came out of the water and then almost effortlessly and silently disappeared below the surface. My heart jumped. Almost two full minutes would pass before I would see the head.

Steve Redding, an experienced free diver and our hired gun for this research trip, silently broke the surface near the kelp bed and floated there motionless. He did not use a tank full of air or a regulator to enable him to hunt underwater. He relied solely on his own breath-holding capability to sustain his dive. When I became obsessed with free diving later on, even at my most accomplished, I never came close to staying below the surface without air for two minutes.

As he circled near the boat, I could hear him breathing through his snorkel. Just for fun, I decided to see how long I could hold my breath. I sucked in sharply, closed my mouth and let my cheeks puff out. After about 20 seconds, I felt like I was going to pass out and I could feel the sour taste of my breakfast teetering at the back of my throat. I frantically exhaled in complete failed disgust.

Steve and I had first met on that second research trip. He was a genuine guy that you felt a friendship with as soon as you spoke with him. Since he was an avid and experienced spear fisherman, he was routinely hired to assist on these trips to gather larger specimens of the fish we were studying. And gather them he did. He had explained to me that because he didn't use a tank or a noisy regulator, he was essentially silent underwater. He would use this stealth to sneak up on fish without their ever knowing he was there. His gun was equally as silent and I had observed firsthand just how deadly this combination could be. Steve was essentially an underwater sniper.

As time passed and I got to know Steve better, he would routinely stop by the fish lab at the university and ask us about the research. He would also bring in pictures of large fish he'd speared at some secret location. The photos almost always pictured Steve and his trophy against some indeterminate background, so you could never figure out exactly where he was. Whenever I'd ask about the location, he'd always just smile wryly and promise to take me there someday.

Steve remained still in the water for a few more minutes, slowly drifting the edge of the kelp bed. He then peeled away from the kelp and swam towards the boat. His three-foot-long swim fins almost effortlessly propelling him. He surfaced right in front of me and waved. "Did you get anything?" I asked. He spit out his snorkel and lifted his mask. "There's nothing big here," he responded. Earlier, Steve had shown me a photo of a white sea bass over fifty pounds he had just recently shot at his secret spot, and I had to wonder what this guy considered big.

I reached down and helped him with his spear gun and stringer. He lifted himself out of the water and sat on the narrow swim step and popped off his mask. "I heard the tank divers get in" he said, looking out over the kelp. Apparently trying to be quiet in the water with a tank on your back is pointless.

I told Steve about the abalone iron Lisa had dropped overboard. "She dropped it right here?" he asked, pointing next to the boat. I nodded and carefully leaned his spear gun on the transom. I watched as Steve put his mask back on and slipped back into the water. He looked up at me with a grin and said, "I'll be right back." Before I could mention the depth, Steve was gone. I watched in amazement as his dark shape headed straight down and disappeared from view some thirty feet below the surface. I believed there was no way he would reach the bottom.

A full minute went by and I started to become concerned. I stared down into the depths, squinting, hoping to see any signs of Steve returning. I turned to look for assistance on the boat and saw no one. I looked back into the water. I saw the bright flash of yellow first. The blurred color seemed to float a short distance directly in front of the returning diver. When he neared the surface, I realized that he was holding the new iron in his teeth.

I shook my head in disbelief as he broke the surface. "Is this the one?"

I was speechless. He handed me the iron and lifted himself back on board. "Man, that was deep," he said. The ship we were on was 85 feet in length, and Steve had gone another twenty feet beyond that length to retrieve the iron. Back then I don't think I could've walked through the galley holding my breath. To this day, I think how that dive was one of the most amazing examples of human endurance I had ever witnessed.

In the years that followed, Steve became a regular fixture on our research trips. He was as willing to share his knowledge of the sport of free diving as he was to learn about the species we chased and the research we were conducting. I was even fortunate enough to dive with Steve on a few occasions during my years on the boat.

Following that first encounter with Steve, over the years I spent a lot of time in the water learning all I could about the sport of free diving. Early on, I shot whatever was in range. Once I became more experienced, I learned to target bigger fish as well as the more edible species. And once I got comfortable, I spent every spare moment in the water diving, chasing fish, and learning everything I could about the sea and what lived there.

One Saturday, I returned home after a successful dive in a great mood. I had speared a large halibut and was excited to fillet the fish and cook it for dinner. Back then, I could think of nothing more satisfying than heading into the sea to hunt and returning with wild food.

After I cleaned up, I noticed someone had called. The only message was from my friend Carrie and she sounded distraught. After a short conversation, I hung up the phone and felt numb. I glanced over to the still dripping spear gun leaning in the corner, and I slowly lay back on the bed. My friend Steve was gone.

He had been diving with a group of divers at one of the Channel Islands. As was frequently his pattern, he had chosen to dive alone in search of big fish. When all had returned to the boat at the end of the dive, Steve was absent. A quick and frantic search of the area found him in 100 feet of water, motionless and likely the victim of shallow water blackout. Attempts at resuscitation failed.

Shallow water blackout is the dark menace of the free diver. When a diver sucks in a lung full of air at the surface and heads to the bottom, his body begins to compress. If he spends too much time at the bottom, his body will have removed practically all the oxygen from his compressed lungs. As he resurfaces and the compression is eased, the now oxygen-starved lungs expand and begin to suck oxygen from the diver's blood. Unconsciousness occurs due to oxygen starvation and usually hits the diver only 10 to 15 feet below the surface. If no one's around to assist you out of the water, it is the end and at the bottom is where they'll find you.

After the call, I lay there for a while mourning the loss of a truly great guy and a friend. I thought about the day I met Steve and the times he'd stop by the fish lab to just visit. I used to think that Steve dived at depths where only he and fish were comfortable. I'm almost certain that he never knew the influence he had on me in the relatively short period we knew each other. Later that evening as I ate the halibut I speared, I knew the only reason it was sitting on my plate was because of what Steve had taught me.

The next morning I rose early, loaded up my dive gear and headed out to the beach alone. I didn't feel excited to dive, but I wanted to wash away the sadness and I thought it would be a fitting tribute. Once I got to the beach, I suited up and walked into the surf. The sun was just peeking over the horizon and it hadn't had time to wash the gray morning gloom from the sea. I remember feeling sad as I pushed through the waves.

Out past the breakers, I adjusted my mask, pulled the heavy bands back on my spear gun and loaded it. I floated there motionless at the surface for a few minutes, letting my eyes adjust to the light and my body to the cold. I pointed myself toward the kelp and slowly headed for the reef a short distance offshore to hunt. After a few minutes of kicking, I found myself floating over some familiar rock formations on the bottom. I drifted there, searching the sandy floor in the low light for fish. At an open spot, I took a few deep breaths and then headed towards the bottom.

I spent the next hour diving to the rocky bottom in 25 feet of water and weaving my way through the kelp bed. The visibility was great that day and it seemed like I could hold my breath just a little longer than usual. Using some of the underwater hunting tips I had learned from Steve, I was able to sneak up and get close to several large fish. But my mind was elsewhere and I never took a shot.

On the last dive of the day, I dropped down onto a sandy area that was bordered on one side by a deep, undercut ledge. I lay on the bottom, grabbing the rock to hold myself in position as my eyes adjusted to the dark cave under the rock. The yellow eye of a five-pound sand bass appeared first as it floated out towards me, curious yet cautious. I slowly swung my spear gun into position. I felt my finger slide to the trigger and then just as quickly slide back off. I could feel it; I wasn't a hunter that day and I let the fish be.

We stared at each other for a few seconds and part of me never wanted to go back to the surface. The nervous fish switched its glance between me and the spear tip. That morning I was just happy for the encounter. I pushed off the rocks and slowly kicked for the surface. The dive was over.

Back at the beach, I dropped my gear and sat in the warm sand. I stared out at the calm sea and watched the kelp swirl with the current. It had been a good dive. I then wondered how the same ocean that I had just · enjoyed could also take my friend Steve. I stared at the shore and become hypnotized by the cycle of the waves. I felt the sadness, and for the first time since changing my career path, I realized that this was real. That when you choose a life in the outdoors, nature calls the shots and there are no rules. Unlike engineering, the world of the ocean sciences can't be drawn up, measured, and manufactured. And except for the relatively predictable tides, she does what she wants, when she wants. She does not care that I study her or her inhabitants, and she would just as soon add me to the unseen clutter at the bottom, as present me with perfect conditions for a dive.

I unzip the top of my wetsuit and push the heavy rubber off my shoulders. I struggle briefly to extract my arms from the foam tubes and with a satisfying pop they pull free. The breeze dries the sweat and seawater from my back and it feels good. I look down the long shoreline and my eyes follow it until it bends inland and out of sight. To me, the shore has always been much more than a line that separates land and sea. It is a boundary; and in its most primal role, it delineates the tamed from the untamed.

The affair between man and the sea has always been a tumultuous one. As soon as we developed the courage and the knowledge to leave the shore, the sea has been there patiently waiting. She has swallowed the prepared and the unprepared with equal indifference. She has sunk the unsinkable and has made the hopeful label "missing at sea" more often synonymous with death. And on any given day she can snatch innocents from the shore of a sunny beach, and spit them out face down and bloated. In short, she does not care. She is a living, breathing, modern-day monster and she sees me as nothing more than an insignificant sack of nutrients that would barely be worth her effort to absorb.

I toss a shell into the surf and watch it disappear forever. I dig my bare heels into the wet sand and smell the familiar scent of the sea. I think of a dimly lit computer lab that was part of my old life and wonder how I tolerated it for so long. A crashing wave thunders over the sand and like a spoiled child demanding my attention, slams the shore loudly, almost as if to say this is your life now. I have given up brain-numbing boredom for the wild and unbridled sea. I thought about my friend Steve and I felt my eyes well up slightly. I tried to shake the sadness away, but it didn't work.

INTO THE SEA

A few months after that second trip on the *Yellowfin*, I had an opportunity to try free diving for the first time. The recent research trips had made it clear that if I was to be serious about my field biology training, I needed to get into the water. The onboard work was interesting, but the direction of the program would eventually lead me to the edge of the swim step and a whole different world beneath the surface. On the last day of the trip, I discovered that absolutely everyone on the vessel, with the exception of me, was either scuba certified or comfortable in the water. Minus any snorkeling gear, I was left alone on deck as the rest of the crew enjoyed an island dive to end the trip.

Before I ever got into the water, I can remember having a real apprehension that bordered on genuine fear of swimming in the ocean. At that time, I had done very little snorkeling and absolutely no diving, so I had no idea what to expect. That fear came from growing up in the shadows of such movies as Jaws and The Beast. Even into young adulthood, unimaginable creatures with huge teeth and strangling tentacles waited just beyond the waves in my mind. It makes you wonder how many potentially brilliant young marine scientists were scared senseless from the shore by the novels of Peter Benchley.

I can remember very clearly that first dive. I didn't sleep much the night before and twice during my early morning breakfast I almost vomited. I know now that my excitement regarding my new career path had a lot to do with me pushing through that fear and conquering it. I was absolutely committed to learning to dive. And even though my first dive wouldn't go quite as smoothly as I'd liked, I would quickly become more comfortable and at ease in the water as time went by.

Tom Grothues, an experienced free diver, had been on that same trip with Steve and had volunteered to take me out for the first time. The morning of the dive arrived dark, cold, and very early. I rubbed my hands nervously waiting for Tom to pick me up. As I stood near the front of my apartment building, the last place I wanted to be at that hour was floating in the ocean.

I heard Tom's late model Datsun before I saw it. The rattling valves and oil smoke demonstrating that the rusty station wagon's days were numbered. With a piercing metal on metal screech, the gray vehicle pulled to a stop in front of me. Tom leaned over to the passenger side and motioned for me to toss my gear into the back. After slamming the hatch several times to get it to closed, I climbed in. It wasn't quite 5:00 AM.

Tom's car was definitely more about function than form. The interior smelled of mold and seaweed. The floorboard, on my side, was more sand than carpet and an old mesh dive bag was tucked under the seat. Two large spear guns were haphazardly tossed near the emergency brake, their exposed metal tips scratching the dashboard as we drove. Dirty dive gear was strewn in the back, most covered in dried sand. Tom's storage locker on wheels was more about getting us to the beach and less about looking good.

I was a little surprised to see two spear guns in the car. I figured that a first dive would be focused on getting comfortable with all the gear and the ocean. Tom, apparently felt differently. The long teak and stainless steel weapons looked far nicer than anything you would take into the ocean. The trio of thick, caramel-colored surgical bands at the end of the gun looked short and strong. I wondered if they would stretch back far enough to load the weapon. Both guns were Riffe Standards and retailed for about $450.00 or about double what I made in a month. Tom saw me looking at the spears. "Don't worry, I'll show you all you need to know when we get to the beach."

At about 6:00 AM we drove into a vacant beach parking lot and down near the shore. Jeff and Chris, fellow divers, were already parked near the surf enjoying the comforts of a working car heater. We all got out, greeted each other and then stared out towards the shore. The ocean looked truly angry. The waves slammed the shoreline with thunderous explosions of green-tinted water. The sun was just peaking over the horizon and gave the scene a slightly more inviting glow. However, the ocean looked like it didn't want us there.

A van pulled up in the parking lot and several surfers got out, boards in hand. I knew enough about ocean conditions that if surfers were out and ready to enjoy a south swell, divers should probably stay home. The wave conditions that bring surfers out, churn up the water absolutely destroying the diving visibility. Tom continued to stare at the surf for several minutes.

I had already gotten back into the car. Nothing about the ocean looked inviting to me and I was content to let my first dive wait for another day.

I watched the three discuss, shrug and point towards the beach. I was relieved to see that they were also having second thoughts. Tom said something, pointed south and they broke up and headed back to the vehicles. "We're going to try someplace else." I had already convinced myself that we were done for the day and should head to Denny's for all you can eat pancakes. As we got back on the road, my nervousness returned.

We traveled south, Tom leading the way. The next spot was a bluff far above the beach. We got out and traveled a well worn path to an overlook that gave us a great view of the surf. The white caps just off shore turned the ocean gray and the waves continued to beat the shore with green water. To me, this spot looked even rougher than the previous one. I stood on the bluff behind the others almost willing them to call it a day. I realized that learning to dive was in my future and eventually I'd get it done. What I didn't want to have happen was to force the issue and have my first dive be in questionable conditions.

Back in the cars, we headed further south in silence. We drove through a residential area near Malibu and stopped at a public access gate. Two large signs bolted to the brick columns read, "Private Property, Keep Out". We got out of the cars and walked down a flight of stairs to the sand. Tom tapped the sign as he passed. "These people like to think they own the beach." We walked through the unlocked gate between two large beach houses and found the ocean. It was like looking at a swimming pool. The waves were small and subdued, and the water out as far as I could see was clear. The white caps were gone and the ocean's surface was what I would call calm. There wasn't a surfer in sight.

We dragged all our gear down to the sand and squeezed into our wetsuits. Even though I had practiced putting on the heavy, rubber suit the day before, my unfamiliarity with my gear had me lagging behind the others. Tom walked over with a single spear gun and started talking before I looked up. He handed me the gun, which felt a lot heavier than it looked, and gave me some brief operational advice. "Don't load it until you get into the water and don't get out of the water with it loaded," he said. After a few minutes of instruction, he walked over and grabbed his gear and scooted into the surf, leaving me on the beach alone. Just before he turned to head for the kelp beds off shore, he gave me one more piece of advice. "Don't lose it!" I watched as Tom pushed passed the waves, adjust his mask and then kick towards the kelp bed a short distance from shore. Jeff and Chris were already in the water and near the kelp beds themselves. I looked down at the gun and then back out towards the departing divers. Apparently the first lesson in taking a new guy out for his first dive was every man for themselves.

Cradling the gun, I put on my brand new $50.00 weight belt and grabbed my fins. I adjusted my mask and scooted into the surf, side-stepping like I had seen Tom do. Just out beyond the waves I turned towards the kelp bed, put my snorkel in my mouth and dipped my head into the water. As soon as I did, cold water flashed over my face and down the back of my wet suit, instantly taking my breath away. I reared up, gasping for air, frantically paddling in five feet of water. Feeling the bottom, I closed my eyes and told myself to relax. I regained my composure, tightened my grip on the wooden weapon and began slowly kicking towards the kelp.

Moving away from shore, the sandy bottom dropped from view. My breath was still hurried and spastic as I fought through the mammalian paradox of being able to breathe with my face in the water. I slowly kept kicking, trying to control my breathing. Every few minutes, to make sure I was headed in the right direction, I would lift my head and adjust my course. Shortly after I left the shore, I noticed that the others had reached the kelp bed to hunt.

After a few minutes of kicking, I decided to just float there and rest. I closed my eyes and tried to get used to all the gear and my new surroundings. One friend had told me that the first dive was always the toughest and it may take a while for me to feel completely comfortable. Some of my more eclectic friends had even assured me that once I got used to the freedom of the ocean, I would eventually feel relaxed and a bit euphoric. As I floated there, I tried to think of what would pass for euphoria. Visions of my effortlessly cavorting with Flipper seriously scared the hell out of me. And I can say with a most convincing certainty that if I had seen any sign of a dolphin in the murky water that day, it would have been in danger of getting shot out of my sheer reactive terror.

After resting, I continued to kick for the kelp. Progress was very slow and I could feel my heart pounding in my chest at the effort. I sucked huge gulps of air through my snorkel and I was definitely expending a great deal of energy without seeing much of a benefit. My fins were leaving the water with every kick, and for some reason, traveling in a straight line was out of the question. With no visual reference in the water, I had to keep correcting back to the right after veering off course. No matter what adjustments I'd make, I could not swim in a straight line.

Thrashing about, I started wondering what the problem was. I'm right handed, thus right footed, and possibly my stronger right leg was causing my direction problems. There wasn't much I could do about that. I then began to wonder if maybe some unnoticed terrestrial defect, something that didn't matter on land, was being magnified in the unforgiving ocean. I had no answers. My zigzag path to the hunting grounds was not only inefficient; it was also wearing me out. I took another rest and glanced up to see where everyone was. All three of my fellow divers were now swimming among the kelp, about 100 yards from where I currently floated.

It had been about twenty minutes since I left the safety of the shore. During my swim, I had been awkwardly shuffling the huge spear gun from one hand to the other. The loose bands would shudder in the water and make this strange humming sound, sending a vibration down the wooden stock. I held the gun in both hands, staring at it. I decided it was a good time to load it. I placed the butt of the gun on my hip like Tom had shown me. I reached up and grabbed one of the thick surgical bands. I pulled the band towards me with both hands and attempted to load it to the spear. For a few seconds I shook slightly as dense rubber and muscle battled. I was no longer a lean, floating machine. As I struggled at this new angle, I began to tip sideways and water washed down my snorkel. Distracted, I let up a bit and the bands began to pull back. Without warning the gun slipped off my hip and sailed almost unobstructed through my legs. Euphoria was not the feeling.

I didn't realize it at the time but my inexperience with the weapon had proven both painful and costly. The butt of the gun had not only slapped my man parts, but it had also caught the release buckle on my brand new weight belt, sending it straight to the bottom of the sea.

The pain was unbelievable. I rested there curled up in a floating ball. Everything from my lower back to my eyeballs throbbed and ached. I rested for a few minutes, and with some considerable courage on my part, I was able to load two of the bands on the spear gun. I took a deep breath through the snorkel, and again pointed myself towards the elusive kelp.

The gun accident had left me with some very strained stomach muscles, making swimming painful. I could also feel that my legs had lost some of their determination and strength. The missing weight belt was also complicating my efforts. Without it, my buoyant suit kept me floating more on top of the water, rather than in it. I did notice that my swim fins were leaving the water more as I kicked after I loaded the gun, but I wouldn't find out I lost the expensive weight belt until later.

Since leaving the shore, I had seen absolutely nothing in the water. The sight of the long, brown stalks of kelp off to my right at first startled me. It had taken me about 20 minutes to swim about 100 yards. Most of that time was spent correcting course and recovering from self induced pain. Once I got to the kelp bed, I grabbed a hold of one of the plants and rested. I was relieved to be with the rest of the group, but I didn't feel very good. I closed my eyes and tried to relax. At this point I was exhausted. Without a weight belt to keep me stable, the swell was tossing me around like a cork. I could feel my handle to the kelp being strained and tugged, and I didn't feel at all comfortable anchoring myself in one spot. I released my hold on the brown plant and floated deeper into the kelp bed. Almost immediately I realized that I was done. My mouth was dry and my head hurt. I felt dizzy and while my stomach muscles felt better, my stomach itself did not. I could feel the dark cloud of nausea settle over me. I was getting sea sick.

I tried to swallow hard to settle my stomach. It didn't work. I lifted my head and spit out my snorkel to get some fresh air. As I took a breath, a few ounces of seawater washed into my mouth and down my throat. I gagged strongly and coughed up what I could. I took a couple of really deep breaths and pushed the nausea out of my mind.

I floated around the kelp bed on auto pilot. I lazily kicked, drifting through gaps in the kelp, looking for fish and trying desperately to feel better. By now the rest of the group had split up at the hunting spot and could not have been less concerned with where I was or if I was still alive. The water was murky and the plants floated eerily at the edge of visibility. For some reason having something to actually look at slightly settled my stomach.

Having never used any of my gear in water previous, I had no experience of what you should and shouldn't do while diving. Earlier, while adjusting my mask, I had actually popped the seal to my face to adjust it. Having broken the seal, water now slowly entered the left side of the mask continuously. No matter what I did, I couldn't get it to stop. Towards the last part of the dive, I had to keep my left eye shut to keep it from being doused with water.

While I was messing with the mask, I drifted into a large kelp paddy near the center of the kelp bed. The long brown stalks and greasy blades seemed to coil around my body. At first I just brushed them away, believing I would drift right over the mat. But the swell pushed me further into the center and I started to get tangled. After a few more moments of struggling, I became covered with the greasy weed.

So there I was. The ocean had carefully guided me to the center of the kelp bed and jammed me into the throat of the kelp. And whatever gear I still had left, didn't appear to be working properly. My mask now leaked like a sieve and I had given up trying to fix it. Unfamiliar with the simplicity of swim fins, I was convinced that I had possibly put them on the wrong feet. I also blamed my failing gear on my current physical condition. While my stomach muscles did feel better, I marked the groin injury as the beginning of the downslide of this dive. Pushing the nausea to the back of my mind seemed to only prolong the inevitable. While stuck in the kelp, I conceded defeat and knew soon I would be vomiting in the sea. The cherry on top was being wrapped in a greasy plant and anchored to the sea floor like some sacrificial offering. At this point, I would have to say that the dive was not going well. The single bright spot of my situation was that I still had Tom's gun.

The slimy plant wrapped around my face and slid across my useless mask. Every time I tried to pull away, sea water doused my snorkel and drained into my mouth. Each inhale was followed by a powerful exhale, to purge my snorkel of water and to grab another breath. Exhausted and lightheaded, I focused all effort on my snorkel, keeping that flimsy tube that supplied me with oxygen, above water and pointed skyward.

I spent a few minutes using Tom's gun to battle the evil weed and actually extracted myself from the tangle. I floated there completely spent next to the kelp ball. I had absolutely no desire to continue. I had arrived at the beach that day to push through my fear and to become a free diver. At this point I was no where close to the level of diver. I had littered the sea floor with gear and I had injured myself in the process. Considering my crippled mode of travel, the only title I was qualified for was free floater.

I straightened and drained my mask, blinking the seawater out of my eyes. As I cleared my vision, I glance back towards shore. The distance I had come astounded me. I was now easily 300 yards from the beach, and the calm conditions that had greeted us earlier had degraded. White caps started to form and the swell I had been fighting during my trip out was now slamming the coast with large waves. The scene looked daunting, but I think at that point I would've walked through fire to get back to land.

Back out towards the kelp, I spotted another diver close by. Even in considerable misery, I was beyond elated to see another person out there. Tom spotted me and motioned me over with several excited hand gestures. Every cell in my body wanted to return back to land where mammals belong, but I limped over near Tom, giving every kelp plant a wide berth.

Using one hand, I dog paddled over, splashing heavily. I had abandoned any attempt at stealth. As I got close, Tom turned around pointing down. "Hey, there are a lot of –" Tom started. He stopped in mid-sentence and just looked at me in silence. "Where's your weight belt?" he asked. I reached to my waist and searched in vain for the expensive piece of equipment. I knew that my physical pain and discomfort would pass quickly, but the loss of my expensive belt had put a monetary penalty on the day. At that point, I regretted ever leaving the shore. I closed my soggy eyes in frustration and did what I did best, floated.

With very little persuading, Tom recommended that I head back to the beach. I turned my tired body towards land and weakly kicked for shore. I was done. I had no desire to work through the nausea or the pain. I just wanted to find a warm spot on the beach and lay down.

Switching direction and heading back, I began to be pushed around more by the developing swell. This made my head hurt and further upset my stomach. When I made the decision that the dive was over and to turn around, I was hopeful that I could make it back to land without swimming through a cloud of my own vomit. Now as the waves tipped me from side to side, it appeared that I had no control over the situation or the eventual outcome. I had dealt with sea sickness before and it always seemed to effect me more mentally than physically. I'd always try and push back the dull squeeze of nausea to avoid vomiting, convinced that leaning over the rail made me look weak and less manly. This psychological fear had always effected me far more than the physical relief I would feel after I horked over the side.

I kept moving. I tried to swallow but my mouth was beyond dry. I was surrounded by water and you could've easily struck a match on my tongue. I could feel that I was on auto pilot. My legs worked weakly without much instruction from me. The front of my body was on air patrol, blowing the sea water out of the snorkel that seemed to wash in with every swell. My arms dangled useless at my side, used now only to hold and protect Tom's gun on the journey home. The nausea was debilitating and my tired body no longer had the strength to fight what was coming.

I spit the snorkel out just in time. My body convulsed and I spewed out a cloud with surprising velocity. The whole and partially chewed Cheerios shot from my mouth to freedom and popped to the surface. For a few seconds I floated there completely spent, drifting near my recycled breakfast. I was perfectly content to let the tide take me in and wash me up on the beach like a discarded piece of trash. After a few minutes of floating, I looked towards shore and started kicking for the beach.

Within a few minutes I felt the ground swell slowly lift me towards shore. I was so exhausted I just floated there and waited for the swell to move my tired body towards land. I really thought that the worst was just about over.

Every single aspect of the dive, with the exception of how to use Tom's gun, I was working through on my own. I had been given no advice on how to use the gear or tips on what I should do once I entered the water. The total training for my first dive consisted of driving me to the beach and pointing to the kelp off shore. I was definitely winging it. I figured since we had simply walked into the ocean at the beginning of the dive, that we'd just walk back out when we were done. I had absolutely no concept of timing my exit between the larger sets or waiting for a lull in the surf. I just forged ahead like I was storming the beach at Normandy.

I began to think something may be wrong when I first saw the ocean bottom in about ten feet of water. Despite my closeness to shore, I could feel myself being pulled backwards. I remember feeling a large presence behind me as I was pulled up the face of a monster wave. The mountain of water crested and engulfed me completely. I was no longer in control. I wish I could say that survival took over at that point, but it became pretty clear to me that if I was part of a clan of primitive water people, natural selection would have snuffed me out right there.

The wave crashed down on my back and pushed me straight to the bottom. My chest scraped briefly along the sandy floor, and then I was tossed backwards into the churning surf. In the chaos, the mouthpiece of my snorkel was ripped from my mouth. My mask was doing nothing more than holding about 12 ounces of water against my eyes and I was so contorted during the exit that I felt one of my useless swim fins slap my back. I had lost all sense of self. I was a meaningless object in the surf and

I had absolutely no say in my direction or whether I would survive or not. I did make one conscious decision. With total disregard for my own safety, I made sure that nothing happened to Tom's gun.

After what seemed like minutes, I finally righted myself and pushed off the bottom towards the surface. I didn't have to go far. As soon as my head broke free of the water, I pulled in a huge lung full of sweet air. Breathing never felt so good. Once at the surface, I realized I could stand. The water was about chest high, but it rapidly started to recede. Once it got down to my legs, I lost my balance and landed on my knees. Seconds later I was on all fours in about 12 inches of water. Clear as day I can remember thinking that I was not too proud to crawl. I don't even remember moving from that spot.

I was right in the danger zone, in the most submissive posture imaginable. It was like I was waiting for the ocean to give me a spanking. The immense amount of water that hit me caused me total confusion. I was instantly pinned to the sandy bottom by a weight that squeezed the air from my lungs. Once again I was tossed almost weightless into the churning surf. Tom's gun, which I had made certain to protect during the first assault, had turned on me. The butt of the gun slammed into my stomach, slid up my chest and whacked me in the chin. My mask and snorkel had been pushed off my face and now both hung useless around my neck. As I was rolled uncontrollably around the beach, I realized that I wasn't having trouble "exiting" as experienced divers put it; the ocean was simply kicking my ass.

This time the wave graciously deposited me scant yards from the comfort of the beach. I stared at the wet sand, gasping for breath on all fours. As I blinked the saltwater from my eyes, one of my swim fins washed down the beach and landed near my hands. I grabbed the useless piece of rubber and scooted out of the danger zone. I dropped all the gear I still had left above the tide line, collapsed on the dry sand and waited for death.

To this day I'm almost certain I either fell asleep or passed out there in the warm sand. My jittery equilibrium was so rattled from the ordeal, that I was convinced I could actually feel the Earth's rotation. The sun beamed down on my tired body and never felt so good. I could've stayed there motionless forever.

I have no idea how long I laid there. With my eyes closed, I could hear a fellow diver sloshing through the surf towards me. I must've been quite a sight. I wasn't simply relaxing on the beach enjoying the warmth of the sun. I was spread eagle, tangled with gear and kelp, covered with drying sand from head to toe. It looked like the ocean had chewed on me for a bit and then spit me out.

Tom's disappointed voice shattered the tranquil silence. "Are you all right?" he asked, not at all trying to suppress slight laughter. His concern was touching. I slowly raised my head and looked at Tom. I'm sure he was taking a visual inventory of the gear he had lent me, especially the gun. I had followed the limited rules he had given me about the weapon. I hadn't loaded it until I was in the water, I didn't lose it, and the ocean had unloaded it for me during the second spin cycle.

Tom looked the exact opposite of me. First of all he was standing. He did not have a grain of sand anywhere on his wetsuit. His mask, snorkel and fins were threaded on the barrel of his spear gun, and his yellow tag line was neatly wound around his gun and tied off in some unfamiliar sailor's knot. Hanging from his weight belt was a fish stringer holding a large sargo, its pectoral fins gently swaying in the breeze. Even the dead fish looked disappointed. Tom could've been posing for the cover of some free diving magazine. The way I looked, I could've made the back page of a clean-up-the-beach brochure.

I told Tom I was fine and I'd meet him back at the car. I lay my head back down and closed my eyes. I thought about one of my dad's old pilot mottos; any landing you walk away from is a good landing. No matter how I twisted the phrase, I couldn't make it work for diving. I had survived the ordeal, but it wasn't pleasant. If I had been asked at that very moment about the dive, I would've stated that it was a complete failure. But I no longer look at that way. I learned a great deal of what not to do during that dive and some of those lessons I still carry with me. The most important lesson I learned is that you alone are responsible for your own decisions and survival out there.

Three days after that dive, I got back in the water and tried again. This time I was equipped with a Scopolamine patch behind my ear to quell the sea sickness. I borrowed a weight belt and a smaller, cheaper spear gun and headed to the beach with a few friends. After two hours of kicking around the kelp beds chasing fish and absorbing everything I could, I carefully made my way to shore and exited standing. I had not only got back on the horse, I felt like I had won the Kentucky Derby.

YOU WON'T CATCH
ANYTHING HERE

In the spring of 1993, two years after starting my course work at Northridge, I graduated with a Bachelor of Science degree in environmental biology, with an emphasis in fisheries. All the units I had transferred from the night classes enabled me to cut my college time in half. And after spending over eight years pursuing a degree, I finally had it in my hands. To this day, I list getting that diploma as one of my highest accomplishments.

I wish I could say I had a solid career plan for after graduation, but I didn't. To be honest I didn't feel at all prepared to enter the work force. I had spent so many years in school, I had no thought of continuing on or the future. I was definitely in limbo.

To make ends meet, I cleaned up around the fisheries lab and made sure that I was involved with any paying contracts the university had to offer.

Most of the research trips that were conducted through the fisheries program were the result of contracts that had been awarded to the university through proposals submitted by the lead investigator, Dr. Larry Allen. This was not only an excellent way for undergraduates to gain real research experience in fisheries, it also served as a source of income for us starving students. And I can emphatically say when the summer of 1993 rolled around, I was most definitely starving.

The work for these contracts was usually back-breaking, but familiar. The pay was more than we'd get for off-campus jobs, so the decision to come aboard was easy. We also got to spend a great deal of time on research vessels and working in the field sciences as biologists. To me, the work experience was much more valuable than any paycheck.

One of the more lengthy contracts we were involved with was a fish sampling project for the Navy in San Diego Bay. The project involved surveying the fish assemblage inside the bay at several habitat stations. We used the *Yellowfin* as a home base and to sample some of the deeper portions of the bay. The smaller skiffs were used to deploy small purse seine nets, as well as beach seines and beam trawls in the shallows to sample fish. These trips were usually a week or so in length, and while the work was tough, we were usually finished with the day's sampling by lunchtime.

The fisheries work involved five different sample stations within the bay. At each station the work crew would conduct three separate sampling regimes; beach work and shallow water beam trawls, small purse seine and large otter trawls. The two smaller work boats conducted the beach and purse seine sampling. After the small boats were finished, and once the work crew was back aboard the *Yellowfin*, we'd assist with the large otter trawl sampling deployed off the mother ship. Weather and conditions permitting, we'd usually be able to complete one entire station per day.

Each small work boat would have a different task and crew for each station. One work boat would use a large beach seine near the shore and sample the fish species in the shallows. Once the seining replicates were done, the beach crew would sort the samples, stow the seine and bring out the cube of death. A one meter cubed enclosure that had an open top and bottom. The cube was randomly placed in several inches of water and pushed down into the soft sediment to keep anything from escaping. Several ounces of Rotenone, a natural occurring fish poison, would be added to the water inside the enclosure. Several of the crew would use long handled nets to mix the poison into the water. In less than a minute, small gobies and other shallow water fishes would float lifeless to the surface. And that is why we called it the cube of death.

Once the beach work was done, that same work boat would return to the *Yellowfin* to offload the samples and pick up a beam trawl. These small vessel trawls have a steel, rectangular mouth about six feet wide and are easily deployed from the small work boats. The work team would then return to the same station and conduct a series of shallow water trawls just off the beach. Once that work was finished, the beach work for that station was complete.

The second work boat was rigged with a launching mechanism for a medium sized purse seine net. Once they arrived on station, the boat would be run in reverse in a large circle as the net and floats were fed out of the side. As they closed the net circle, the purse string at the bottom would be pulled and closed, thus the name purse seine and the captured sample would be hauled into the boat. A series of three to five separate purse seine sets would be made depending on the size of the sample area. Most definitely some back breaking work.

If the sample station was adjacent to deeper water, large otter trawls would be conducted using the mother ship. Greasy cable would be run through the A-frame at the back of the ship and the net would be towed at a specific speed. Huge scoops of the bottom would be gathered by the net as the wooden, outer doors flared outwards opening the mouth of the sample net as the boat moved forward. After fifteen minutes, the net was brought back aboard and whatever was captured was identified, worked up and then tossed back over the side. A station was considered complete when the beach work, the small purse seine and the otter trawls were finished.

On one summer trip, we had just finished up with the day's survey requirements. The big gray ship was tied up to the dock near Shelter Island inside San Diego Bay. Western gulls and brown pelicans swarmed at the back of the boat, picking off dead fish samples washed overboard after the work was done. We were in the middle of the sampling trip and the rest of the day was ours to enjoy.

After cleaning up, I walked back out on deck. I stretched my aching back and heard a definite crack as I straightened up. I flexed my sore hands and inspected the damage. Large red cuts and a few blisters were spread over the fingers and palms. If it wasn't the sampling gear constantly gouging your hands, pulling spiny fish out of nets and buckets usually took its toll as well.

John was stretched out on the back deck warming up in the sun. He had his shirt over his head, but his skinny, lanky frame clearly identified the newest member of the work crew. John and I first met in an organic chemistry class my senior year in college. Dressed in his ROTC uniform, John would walk into the early morning class, find a seat in the back and fall asleep with a pencil in his hand. He liked to bill himself as 135 pounds of oriental fury and I often thought he overestimated that weight by about 20 pounds. He was also the only Korean I knew who listened to heavy metal music religiously, and took great pride in proving through the presentation of his driver's license that his name was indeed John Smith.

I grabbed a lawn chair and sat down next to John in the sun. After the early morning starting time for sampling and being constantly wet and cold, sitting in the sun felt good. We dried out for a bit and then decided to take one of the work boats out and do some fishing. John was new to the fisheries program and relatively new to fishing. His new enthusiasm for both the program and the sport was contagious and it was always interesting having him aboard.

We started gathering gear and rummaging through the school's fishing tackle supply. I grabbed the boat keys and headed for the beaten work boat tied to the A-frame of the mother ship. I glanced down into the little messy vessel floating a few yards off the stern. Several buckets and nets were strewn around the driving area and a slightly rusted, 6-foot beam trawl sat wedged between the center console and the side of the boat. The green trawl net was draped over the beam and partially hung in the water. Being part of the student work crew, we often had to maneuver around the work gear when we used the same boats for pleasure.

I tugged on the worn line, pulling the Whaler to the swim step of the big boat. As I stepped inside the skiff, the stainless steel rail of the beam trawl stripped a healthy chunk of skin off my shin. I nudged the rail aside and cleared the other sampling gear away from the driving area. I swiped the blood on the beach seine on purpose, as I bent down to turn on the battery. Still warmed up from the day's work, the little Mariner outboard roared to life on the first turn. As the engine idled, John appeared at the side and tossed his beat up gear and a few bags of rubber lures into the back of the boat. He then climbed the stairs to where the old line held the little boat to the big boat like a faded umbilical cord. He looked my way and I nodded. In seconds he untied us, tossed the heavy rope into the bow and jumped aboard.

We pulled away from the mother ship and headed for the Coronado Bridge inside the bay. We had drifted the huge cement supports of the large traffic bridge many times before and had always caught plenty of fish. As we ran, John rummaged through the large bag of school gear he had grabbed from the big boat. He held up a large bag of green, curly tail lures for my inspection. They were the lures of choice for this bay and they'd be all we needed.

On our way to the bridge, we past right by the station we had sampled earlier that day. The sample station had not gone well and we ended up having to go back out after lunch to finish up. The beach seine sampling had been fowled by jagged rocks in the shallows, making net pulling and walking difficult. Once we finally finished the beach work, returning to the same beach to sample with the beam trawl was no easier. On the first tow the net end had become untied, or was never tied off in the first place, sending the sample right out the back of the trawl. We had to repeat that sample. The third tow became loaded almost immediately after we began. Since the net was already sampling, we had to continue towing it for the full fifteen minutes before we could clear whatever was dragging the net down. It took another thirty minutes to pull the one hundred pound bat ray out of the narrow mesh and to repair the long slice his three inch barb had cut through the netting.

The purse seine crew didn't have it any easier. During the summer the bay can become absolutely choked with top smelt, a silvery bait fish. We heard over the radio that all three of the purse seine samples had been loaded with the small fish. After hearing that each sample weighed close to a ton, I was glad I had been assigned to the beach crew.

When we approached the bridge, I shut the engine down. We drifted close to the large cement supports and I could see that the current was ripping around the base of the pilings. This was good news. When the water moves, the predatory fishes can sit in the calmer water behind the cement pillars and wait for the smaller bait fish to get swept by. Using minimal energy, the larger fish can dart in quickly, grab their prey and swim back to their waiting area. If you knew where to toss your lure, you could hook up with just about every cast.

I ran the boat up-current and cut the engine. The current slowly moved the boat past the supports and we spent the better part of an hour catching and releasing spotted and barred sand bass. After being chewed on by a few fish, the lures we used began to tear and become useless. Dozens littered the deck of the boat as we tore them loose and replaced them with new ones. The fish were biting and we both lost count of how many we caught. I will say that I'm absolutely sure I caught more than John. After a long day of pulling nets and sloshing through the shallows, I couldn't think of a better way to end a day on the water.

After about an hour, the drift began to slow and the fishing dropped off as well. The current was easing into a slack tide, a lull in water movement between high tide and low. Just as the moving water turned the fish into feeding machines, the slack tide usually turned them off. We sat in the shadow of one of the huge cement supports and cast the lures for show more than anything else. We knew the fish catching was all but over.

We drifted around for about an hour trying to catch anything that might still be hungry. The fish were done feeding and the boat was now motionless in the shadow of the bridge as the slack tide grabbed the bay. I knew during moving water the bridge was a good place to fish. But I also knew of a few others place back towards the *Yellowfin* that may be worth trying. We decided to give it a few more minutes and then head out and find a new spot.

I was at the front of the boat trying to hit one of the warning lights on the nearest support with my lure when I saw something below the surface directly in front of me. At first I thought it was a trash bag swaying in the water about eight feet down. Then I realized that it was alive and about to surface five feet in front of the boat. I took a quick step back from the bow.

As the object neared the surface, I could see it was a diver completely suited in black. Instead of tanks, attached to his back was a round cartridge; he was using a re-breather. This utilizes a lithium scrubber to take the diver's own exhaled air and convert it from carbon dioxide back into breathable oxygen. Because the diver's exhaled air is pushed through the scrubber, he emits no bubbles. No bubbles means no noise, and no noise means no detection. It also gave him the ability to stay underwater for up to 5 hours without surfacing.

The round canister was not the only thing attached to the diver's back. Hanging from the shoulders of the lead diver like a giant tick was another diver also completely clad in black. This diver was using an extra regulator attached to the scrubber to breath, and had no cumbersome gear attached to his back. The lead diver was simply the mode of transportation. Free of any heavy gear and only limited by the length of his regulator hose, the second diver was the underwater assassin.

The lead diver approached the surface slowly. He did not take his eyes off me. He removed the regulator from his mouth underwater. He leaned his head back so only his mouth and chin broke the surface. His own buoyancy control was amazing. If I hadn't seen the pair, their stealthy approach would've been impossible to detect. I stared intently at the diver.

His mouth broke the surface and he began to speak. "You're not going to catch anything here," he said in a low voice. The tick never surfaced. He just peered at me like a freaky little sidekick over the shoulder of the lead diver. The coiled regulator hose at his side meant he could've separated from the lead diver and come up on John's side without our ever knowing. I just froze. I understood the message, but I was confused by the meaning. I didn't say a word, and just stared at the black, two-headed swimmer.

Both men then slowly submerged without a ripple and disappeared into the depths. I looked back at John. "What the hell was that?" I said. John leaned over the side to examine the hull of the boat. He was convinced that they had stuck something to the bottom. He swished his arm in the water along the abused fiberglass feeling for something strange. "You don't suppose they stuck something to the bottom of the Whaler, do you?" he said. I half glanced over the side as well, not completely discarding the idea.

We decided to head back to the big boat. As we ran, John told me all about a program the Navy used to have, training dolphins to plant bombs on ships. Images of Flipper zipping through the surf with a bomb stuck to his head flashed in my mind. I remember thinking how screwed up that was. Taking a relatively peaceful marine mammal and training it for war. That's about as bad as penning them up all over the world, and training them to jump through hoops.

John and I decided that since the Navy Seals training area was close to where we were fishing, we must've drifted over some underwater exercise. The contract we were working on was for the Navy, and they had mentioned that certain areas within our sampling footprint may at times be occupied by military divers. John still wasn't convinced that they hadn't sabotaged our stinky little work boat.

I often wonder what kind of defense I could've mounted against such a sneaky assault. Not many would find a guy in shorts, standing on the bow of a little boat armed with a 6-foot bass rod even remotely threatening. Even my 135 pound Korean sidekick probably brought little more than an amusing smirk to the faces of the underwater assassins. John would bring up the encounter on future trips every now and then, and he still wasn't completely convinced they didn't do something to our boat. When I think about it, I was just glad they spoke English.

DIVING WITH THE FOOD

After the San Diego Bay trip, I found myself once again in the fish lab cleaning up and very deep in thought. I was going through a pretty rough patch in life and was trying to figure out my next step. I was scheduled to assist on one more research trip during the summer, but beyond that the future was foggy. I had no real employment prospects and to be honest, I hadn't given graduate school any thought at all. I had enjoyed assisting with all the research projects during my time in college, but at that point, I felt a little unprepared to go out and get a job. With one research trip left and not even two hundred dollars in my bank account, to stick with the nautical vernacular, I was a ship without a rudder.

I was pushing more dust around than sweeping as I passed Larry's office. He was on the phone and glanced up and nodded when he saw me. I nodded back and just kept on sweeping and daydreaming. I just couldn't shake the feeling that at this point in life I was little more than a janitor that knew way too much about fish. The semester before I had taken an ichthyology class and had really enjoyed the course. During that class I was introduced to the format of scientific writing. All the projects we were conducting in class were to be written up and submitted in the format of a real scientific publication. I can remember getting my first draft back from Larry, actually thinking that his pen had leaked all over my report. I then realized that every mark on the document was a scientific format infraction. My grade for that first project was marginal. I recall having two very powerful epiphanies as I looked down at that document; I wanted very much to be a writer, and I apparently wasn't very good at it.

I was just about to hang up the broom and head out, when Larry leaned out the doorway of his office. "Hey Tim, I have one graduate position available for the fall, do you want to apply for it?" Even though I had taken all the fisheries classes taught by Dr. Allen while at the university, and had volunteered for several research trips, the question caught me by surprise. I had no idea what was involved in applying for graduate school, or what I was required to do once I got there. I can remember thinking that I had absolutely nothing else going on in life during that time and taking the next educational step may not be a bad idea. I leaned against the broom for a second, shrugged and said sure. It happened that fast.

As the idea sunk in, I became more excited and realized that just maybe my janitor days in the lab were over. The position came with a small salary and my graduate status would enable me to teach introductory biology lab classes during the upcoming fall. I can clearly say before that offer, I was feeling pretty low about my future. However, in the span of less than a minute, I started to think that just maybe the bottom had been reached and I had nowhere to go but up.

I put away the broom and decided to head out. Just before I reached the door, Larry called out. "We'll get all the paperwork figured out later and you'll need to sign a university contract at some point," he said. I didn't care. In less than a few minutes, Larry had completely changed my attitude about my future. I think I would've lent him a kidney at that point if he had asked. "Of course your scientific writing is going to need some work," he added with a smile. I already knew that.

With a new enthusiasm about my future, I found myself looking very much forward to the next summer research excursion. The first trip was scheduled for the following month and was part of a contract that I had assisted on before. And while my graduate status was still in administrative limbo and not quite a week old, I couldn't think of any other place I'd rather be.

A few weeks later I was on the back deck of the *Yellowfin* getting the sample gear ready to load onto the Whaler. We were anchored off the coast of Catalina Island and getting things ready to start the day's sampling. This contract involved the collection of larval fish using beam trawls that we dragged behind the skiffs at certain depths around the island. The trawl net had a heavy, rectangular frame about 6 feet wide and about 18 inches high that formed the mouth of the sampler. Attached to the frame and trailing behind the mouth was a triangular mesh net about 15 feet in length. The end of the net or cod end was open so samples could be washed into a sample bucket and sorted back on the *Yellowfin*. Before we started towing the net, we would tie off the cod end so that the sample would be caught at the back of the net during the trawl.

Tom and I tied off the bridle of the sampler to the Whaler and loaded the heavy frame and net into the skiff. I grabbed a clipboard with data sheets and three five-gallon buckets and tossed them into the front of the boat. Tom started the engine to warm up the outboard and I untied the bow line attached to the heavy rail of the *Yellowfin*. I tossed the heavy line into the bow of the skiff and pushed us off.

Tom dropped the throttle to the Whaler and peeled away from the mother ship. The heavy beam trawl net bounced at the back of the boat as we traveled over the swell. We were headed for a stretch of shore just off the coast and about three miles farther north.

A short time later Tom slowed the skiff and began to stare intently at the depth finder. We needed to drop the trawl in 40 feet of water and even though we were close to that depth, we also had to tow the net in a relative straight line for 15 minutes. It was always a good idea to make a practice run through the sample area to make sure we didn't run into any kelp beds or anchored boats.

Tom circled the skiff a few times until he found the right conditions. We lifted the heavy net to the side and as Tom put the boat into gear, I dropped the net into the water. I kept the bridle line tight while Tom slowly increased the speed of the Whaler. Once the line came tight, we started the clock and began the trawl.

We had three samples to collect at three different depths. If everything went as planned, a pair of samplers could get the job done in about an hour and a half.

"O.K., time's up," Tom said, staring at his watch. I had been recording the depth, date and time on the data sheet for the last trawl and I placed the clipboard on the center console. All had gone well for the first two hauls and we already had two samples sitting in buckets in the skiff's bow. Tom slowed the boat as I grabbed the bridle line. He kicked the skiff into reverse and I started pulling the net in. We lifted the unwieldy beam trawl over the side of the Whaler and roughly dropped it on to the deck. The long green catch net was still waving in the ocean current next to the boat. I reached over and grabbed the soaking mesh and started pulling it in. I instantly felt the absence of weight. This meant we either forgot to tie off the cod end, or the sample trawl was termed a water haul and nothing had been collected.

I grabbed the end of the net and found everything in order. "Looks like a water haul," I said. Tom didn't need to hear me say it twice. He moved a few of the sample buckets to the back of the boat and grabbed the steering wheel. He hit the throttle and we raced back to the *Yellowfin* to work up the samples. Once the collected data was recorded and all the sample gear was cleaned up and put away, we were free to enjoy the rest of the day out on the ocean.

Tom was the lead graduate student on this trip and an experienced free diver. I was relatively new to the sport of free diving, but was always anxious to dive whenever I got the chance. When the *Yellowfin* had dropped anchor a short distance from the sample station earlier that day, Tom had mentioned that he knew of some great dive spots in the area. I volunteered to go with him on a dive when the work was done.

Back on board the *Yellowfin*, I could see that the other work boat had already started recording their catch. The total haul for this station looked to be pretty light and so the post-sampling work-up wouldn't take long. Tom apparently saw this too. As I tied up the boat, he leaned over and said, "We'll be diving in no time."

I carried the samples to the sorting table and started recording the data. I could see Jimmy, the captain, pouring over charts in the galley trying to decide where to anchor the boat for the remainder of the day. The student volunteers gathered around the sorting table and started working through the sample buckets. Before we were finished, the rough sound of the twin diesels firing up signaled we were on the move.

We weren't moving long before Jimmy pulled us up to one of his favorite Catalina Island fishing spots. The big gray ship slowly approached the rocky shore and then sharply turned seaward again. The large engines growled as Jimmy slapped both into reverse to steady our position. We bobbed against our own wake, and then the sound of the anchor dropping into the sea meant we were here to stay.

Tom appeared at the sorting table and started working through each bucket we had collected individually. On our first trawl we captured two juvenile white sea bass, the target species for this contract. Each tiny fish was scarcely larger than an apple seed and appeared to have absolutely no control over its own swimming movement. The fish ebbed and flowed with the rocking of the ship in the bright white buckets. I used a small dip net and carefully captured each fish separately. Tom transferred each fish into a small vial filled with seawater and placed small pieces of numbered tape on the lids.

Nothing else of note was found in the first sample trawl. The second sample contained three juvenile kelp bass about two inches in length. Each fish was a perfect miniature replica of an adult and beautifully marked. I recorded the size and species data for each specimen and then tossed the still-live fish back into the ocean. The rest of sample two was also pretty light and nothing else of note was observed. Sample three was an empty bucket, which in the world of beam trawls meant a water haul.

I grabbed some water proof paper and a pair of scissors and walked over to the two vials containing the larval fish. Tom looked at me with a smile and grabbed the paper from me. "I'll label these vials if you don't mind," he said laughing. I just laughed too and handed him the scissors. Both fish were headed for the genetics lab and would eventually meet their demise in a small tissue grinder, but Tom wanted to make sure they made the trip as whole fish this time.

It was on this trip that I began to realize how fascinated I was with the fisheries work. The topics of interest were limitless, and the numerous contracts enabled us to gain experience in everything from species growth rates to migratory patterns. This particular contract involved analyzing the genetics of the larval fish we collected here at the island, and comparing them to individuals collected near the coast. No matter what the project or the species involved, I was interested in learning as much as I could about it all.

As expected the sample work-up for both trawl groups was brief. All the sample gear and data was quickly stored for the next day's collection. Tom and I gathered up our dive gear and started getting ready at the back of the boat. A few of the other students were interested in joining us for the dive. By the time we were suited up and ready to go, we had a full boat of divers ready to explore the island.

The six of us carefully loaded into the same Whaler we had just used to collect fish samples. Tom eased away from the *Yellowfin* and pointed the bow of our Whaler towards the beach. Despite the slight swell, the ocean surface was like glass and I couldn't wait to get in. As we moved past the rocky shoreline, I thought about how it would be nice to spear something big out here. I was relatively new to spear fishing and while I had taken a handful of nice fish, I was still in search of something to truly brag about.

This is how it was for me. While on the mother ship and engaged in the data collecting, I was completely focused on the samples, the protocol and learning absolutely everything I could. I was a sponge and no matter how trivial the detail, I wanted to know about it; almost to the point of being annoying. Once the work was done for the day, a switch flipped inside me and I became a hunter. Since learning to free dive, I wanted to spend every spare moment floating in the ocean looking for fish. I tried to stretch my breath holding endurance on every trip, and add a few more feet to my maximum diving depth. And more than anything, I wanted to shoot a big fish.

We slowly motored near the coastal kelp looking for a good dive spot. Tom looked towards the shore and then dropped the boat into neutral. After a few minutes, he was satisfied with the area and we began gathering our gear. The dive spot was a stretch of coastline about a mile from where the *Yellowfin* was anchored. To spread the group out, Tom decided to drop us off in pairs along the coast. Carrie and I were first. I first met Carrie in my second semester at the university and we quickly became friends. Since we were chasing the same Bachelor's degree at the time, we often found ourselves in the same core classes. She was also a regular on the research trips and having her along was always a pleasure. I consider her the sister I never had.

Carrie and I grabbed our gear and eased over the side. I floated there for a few seconds to adjust my mask. I then kicked back to the skiff to grab my spear gun. As I approached the Whaler I noticed a white line hanging from the boat descending into the depths. The buoys attached to the end of the sinking line looked familiar. They were the ones attached to my spear gun.

Tom had simply tossed my spear gun overboard without bothering to check the depth. Fortunately we were in only 80 feet of water and the buoys were within reach 10 feet below the surface. I took a quick dive, grabbed the submerged buoys and slowly pulled my gun from the depths. I looked up and shot Tom a look of frustration. He just smiled and yelled something that never made it over the engine noise and motored off. I respected just about everything about Tom. To a new graduate student just coming in, Tom's status in the program was to be admired. He was an experienced free diver and good at it. His personality was beyond easygoing and he was always quick with a joke, both clean and dirty. However, at times, he could still be a clown.

Carrie wasn't into spear fishing and agreed to let me lead the way. I slowly kicked over to the kelp bed and drifted along the edge. The water was murky and visibility was between 15 and 20 feet. The kelp looked ghostly as it disappeared into the depths. As my eyes adjusted to the light, I could see dark shadows moving below me, just at the limits of sight. The sunlight danced off the golden leaves of the kelp and then disappeared into the green tint of the water. This was hunting visibility.

As I spent more time free diving, I realized that diving in crystal clear water wasn't the best conditions for hunting. Fish can easily see your approach and will often swim well out of range. Conversely, poor visibility will make seeing potential targets tough. And where vision is only one of the honed senses of fish, once the diver's vision is limited, the chance of success was slim. I learned that the best conditions for me included visibility of about 20 feet, allowing me to make quiet, undetected approaches.

I kicked slowly to the edge of the kelp and floated there motionless. I saw enough activity below me and at the edge of visibility to make a dive. I loaded my spear gun and started hyperventilating slowly. Just as I was about to dip below the surface, I heard Carrie yell something through her snorkel. I looked over and saw her point out in front of us. A pair of 20 pound yellowtail, swimming in perfect unison, glided through the kelp only inches below the surface, just out of range. The fish were here.

I moved through the kelp looking for a good dive spot. When kelp grows to the surface, it doesn't just stop. Once it reaches the top and the life-giving sun for photosynthetic growth, it'll start growing out over the surface of the water forming a shady canopy. If the kelp forest is healthy, many plants can reach the surface and the canopy can become quite thick. Beneath these areas light is blocked out and open, dimly lit rooms are formed. These types of kelp rooms are great for hunting. Unfortunately, the kelp we were swimming through didn't look very healthy at all. We were moving through very thin, single plants that stretched from the depths to the surface and essentially stopped there. They almost looked like terrestrial vines reaching for the sky. There was no canopy for hunting, and essentially no areas to hide.

We kicked through the open kelp easily. The sun and the visibility gave the entire area an eerie glow. You could say it was beautiful and ugly at the same time. It looked like a scene out of a movie where something was just about to happen. I pushed that thought out of my head. In my opinion, in the water is not the place to get reflective and cerebral. Never far from a diver's mind is the thought of something far bigger and more aggressive out there, just beyond the edge of sight. The murky visibility is great for hunting, as it easily hides your approach. However, it can also hide anything that may be approaching you. When I'm free diving, I think about fish and hunting, and not much else. If I started speculating on what else could be out there, I'd never get in the water.

I eased up on a really good looking spot. A stand of kelp formed a wall of sorts that I could use to obscure my dive. On the other side of the kelp wall was a circular opening about 15 feet in diameter. The small gap was bordered on all sides by kelp. Towards the bottom, larger fish floated in and out of view completely unaware of my presence.

I took a few deep breaths and quietly dipped below the surface. I dropped down about 20 feet and made sure I kept the kelp wall between me and the opening. I leveled off and slowly peeked into the gap, my spear gun leading the way. At the far side of the opening and a little below me, I saw the shadow of a fish as it turned broadside. At my approach the fish swam in to see what I was, and then held motionless less than 15 feet from the end of my spear. It was a kelp bass. I raised the gun slowly, adjusted the aim to just behind the head and fired.

I watched as the tag line raced into the depths and disappeared. I grabbed the white line and headed for the surface. Once there, I started slowly pulling in the slack. I could feel the fish struggling at the other end and I knew I had to take it easy. The fish was directly below me but still out of sight in the murky water. I eased in a few more yards and brought the speared fish into view. I grabbed the spear with one hand and the mortally wounded fish in the other.

With my hands full, I tucked my spear gun under my arm. As I floated there, I felt one of my swim fins loosen and then slowly slip off my foot. I looked down and watched as my fin slid off and headed for the bottom.

I quickly dipped below the surface and grabbed the sinking fin with my cluttered hands. I was also able to somehow grab the plastic buckle that had malfunctioned and was making its way to the bottom as well. I re-surfaced with spear, fish, fin and buckle all haphazardly cluttered against my chest. Carrie appeared at my side laughing through her snorkel. She offered up her game bag to hold everything. I loaded it up and then clipped it to my weight belt. I knew I couldn't safely fix my fin while in the water, so I decided to head into the beach for repairs. Carrie decided to stay and explore the coast. The kelp bed was directly off the beach and she wouldn't be hard to find when I returned.

I nodded to her and then headed towards land on one fin. During the swim in, I admired the kelp bass laying length wise and upside down in the bottom of the bag. The deep scar in its belly was white and bits of entrails oozed from the hole. I held the fish in my gloved hands and felt it twitch weakly. Its golden eye looked right at me as I handled it. It wasn't a monster, but it was my biggest kelp bass so far.

My connection with that fish was much more than just something I shot and captured in a game bag. The more I explored the ocean, the more drawn I became to gathering my own food from the sea. And while I was beyond fascinated with this new world and respected it beyond words, at a level more primal than I could understand, I was a hunter. Technology had given me a slight edge, but beyond that, only time separated me from the first off shore hunter. Very little had changed and that's why I liked it.

I turned the bag over and held the fish. By the end of the day I would eat its flesh, its cells meshing with mine and at a molecular level, becoming a part of me.

I sat down on the beach and started repairing my gear. Instantly, I realized that the fins I had left the boat with weren't mine. I turned the fins over and noticed someone else's initials scribed in wax pencil on the back. That really didn't matter now. I unhooked my dive knife and forced the thick rubber strap back through the buckle. I attached the buckle to the fin strap, gave it a good tug and I was back in business.

I gathered up all my gear and made my way to the shore. Before I got into the water, I glanced out towards the kelp bed where I had left Carrie. Off to her right and a little farther out, I noticed another diver floating at the surface. I just thought that one of the other divers had drifted down to our area and I didn't think twice about it. I'm pretty sure that if I had taken a really good look at the other object, I wouldn't have gotten back into the water.

The surf was low on the island and within a minute I was kicking out towards the kelp bed. I reached Carrie hanging out pretty much where I had left her. I also noticed that the other object I had seen from shore was not a fellow diver but appeared to be some sort of lifeless debris floating a few yards further out. Debris that only minutes before was swimming through the kelp bed as free and easy as we were.

Carrie heard my approach and turned around rather quickly. She seemed a bit unnerved and I could see fear in her eyes. She then told me that the debris I had spotted from shore was a half-eaten, freshly killed sea lion. When I heard that, I almost vomited.

I knew exactly what ate sea lions, and I didn't want to be anywhere close to the dinner table when it was eating. It wasn't hard to imagine being injured in the confusion of poor visibility, clouds of blood and the fevered frenzy of huge sharks feeding. Leaving the safety of the boat, we had definitely slipped down the food chain.

I'm not going to lie, I began to panic. I knew handing Carrie the dive bag, containing the bleeding fish, and then pushing off her to get a head start to shore would not have been the polite thing to do. Although I will say that the old survival adage of "I really don't have to run faster than the bear, I just need to run faster than you," did enter my mind briefly.

Together we cautiously began swimming to shore. It's difficult to weigh the urgency of leaving a horrid scene calmly against frothing up the water in sheer flailing hysteria. We wanted desperately to leave the area, but we didn't want to look like wounded prey doing so. We stayed close and calm, and carefully made our way towards the beach.

Within minutes, we were both on shore and out of any danger. Once we caught our breaths, I made an attempt to locate the dead sea lion parts out by the kelp. The object would have been very easy to see in the calm conditions, but we never saw it again.

I sat there looking out at the ocean for a few minutes thinking about the dive. The sea is beyond vast, but the most primitive of scenes had just played out within feet of us. And dressed in black wet suits, we weren't just casually observers in the audience we were unwitting props on the stage. Once you enter the ocean, absolutely nothing matters. Status, personality and education all gets stripped away when you're out there swimming with the food when big creatures were hungry. The order of things changes when you leave the shore.

After what seemed like an eternity, Tom finally showed up in the Whaler to pick us up. We motioned for him to get as close as possible to the shore before we entered the water. We gathered our gear, took deep breaths and then paddled out past the waves. They say the hardest thing to do in diving is to get in the water when you think sharks may be around. I can tell you that this is absolutely not true. The hardest thing in diving is getting back into the water when you know sharks are around.

I kicked as quietly as I could towards the boat and kept my spear gun pointed backwards. My prized bass, now dead in the dive bag, dangled from my weight belt like bait. More than once it bumped my leg as I swam and scared the hell out of me.

Within a few minutes Carrie reached the side of the boat and hoisted herself in. I was still a few yards away but I could hear Carrie start to tell Tom about our encounter. She then stopped suddenly and exclaimed, "Wow!" I had no idea what she was referring to and really I didn't care. I was just focused on reaching the side of the Whaler and getting out of the water.

I was relieved to grab the side of the skiff. I placed my spear gun into the boat and looked up to Tom. He just looked down at me from the center console with the biggest smile on his face. I tossed the game bag over the side and into the boat. The dead fish hit the deck with a satisfying thump. "Hey, nice bass," Tom said.

I lifted myself in and instantly realized why Tom was smiling. Lying in the bottom of the boat, sitting on Tom's swim fins, was a 37 pound white sea bass. I looked over to Tom. He just kept on smiling, slowly nodding his head. The lifeless fish was huge. My fish could've easily fit in its mouth.

Once we got back to the mother ship, the stories of the huge fish and the white shark encounter were told and re-told. I glossed over my heroics since there really weren't any and tried to show off my fish. My story was not half as interesting as Tom's recount of the white sea bass he had speared. I guess he had to dive down to 60 feet to untangle it from the kelp before bringing it to the surface. Since Tom had been chasing a fish of that size for over fourteen years, I happily congratulated him and bowed to the winner.

As the excitement died down, I walked to the back of the boat and looked out towards the stretch of coast we had just been diving. I wondered how close the shark had been. A small splash at the stern caught my attention. An adult sea lion swam by looking at me, interested in the fish scraps tossed overboard as the crew started filleting the day's catch. I watched the animal dart around the ship effortlessly and wondered if it was at all related to the shark victim. I had enjoyed half the dive, but I was most certainly relieved to be back on the boat. It is very clear that as divers we are just observers in the water and we do not matter in that world.

TO THE END AND BACK

One of my first tasks after getting accepted to the graduate program at California State University Northridge, was to decide on a Master's project. The contract that funded part of my research specified that I work on a near-shore marine fish. Dr. Larry Allen, my graduate advisor, strongly suggested I concentrate on the life history of a near-shore species. Since Larry was the boss and my funding dictated the research direction, my choices were narrowed down considerably and made the decision an easy one.

The previous semester, I had conducted a fairly extensive reproductive project on the spotted sand bass for my ichthyology class. I had really enjoyed looking at the reproductive strategies of the fish and figured if I could get Larry's approval, I could broaden the research aspect and populations of this project. With a little convincing, Larry agreed to support my endeavor and actually seemed excited about the research direction.

John Smith and Greg Tranah were also recent graduate students and had yet to finalize their projects. Since every level of the life history of the spotted sand bass needed to be researched or updated, John and Greg agreed to take on other aspects of the same species as their research projects. This would enable us to maximize the samples we collected and easily assist each other with the research.

Since I was interested in reproductive strategies, I would need to collect the gonads and determine the size and age of each specimen we collected. Greg would be working on the geographic distribution of the species and how it related to the physical measurements of each population. He would need blood samples for the genetics work, and he would need to take several specific measurements from each specimen. John would be looking at age and growth of juveniles or young-of-the-year spotted sand bass. A young of the year fish is a fish that is produced in the spring of the same year it was collected. He would need to determine the age of the smallest specimens we collected at each population. Since both John and I were interested in the age of each fish, we would need to remove the ear bones or otoliths from each specimen. Otoliths are small, paired bones located in the auditory sack, just behind the fish's head. These calcified stones could be removed, cut into thin sections and examined under a microscope. The thin sections displayed annual rings that allowed for age determination.

Once we were finished with our individual projects, the combined research would yield a complete evaluation of the life history of the spotted sand bass from a full moon union of egg and sperm to the maximum age and size class of the species.

With the projects firmly outlined, it was time to start making plans to collect samples. The spotted sand bass ranges from Mazatlan, Mexico to Monterey California, including the Gulf of California. Since we were all looking to extend the comparison range of our research, traveling to Mexico to collect specimens seemed not only reasonable but an unexpected benefit of pursuing an advanced degree.

It took us a few weeks to plan and organize the big sampling trip south. Once we settled on the volunteer list, we needed to coordinate university vehicles and boat transport. We used a state van as a food vehicle and figured most of the volunteers could transport themselves across the border in their own vehicles.

The graduate group, with some guidance from Larry, poured over maps of Baja and selected sample locations that would satisfy all of our research goals. The lab had all the gear we would need to collect data and we even had a fishing lure company sponsor part of the trip. All that was left to do now was to head out into the field and collect the samples.

DAY ONE

June 21st, 1993

It seemed appropriate that we were departing for this adventure on the longest day of the year. We met at the university early and began packing vehicles. Tom's noisy Datsun rolled into the compound and screeched to a stop. As soon as he got out, I knew something was wrong.

Tom is a pretty happy person and always seemed to be smiling. As he walked over to the group, he wasn't smiling. He approached the group and explained that while shopping for the trip at midnight the evening before, two men on a motorcycle had mugged him for his food money. We were about to depart on great adventure, and we certainly didn't want the sour taste of crime to taint our moods. We did the best to cheer Tom up. We took up a collection so he could quickly head out and grab some supplies. The packing and small tasks continued until Tom returned. I helped him transfer his gear into Carrie's truck and we headed out.

About an hour after we left the university, we stopped off at the boat yard in Long Beach. We'd be towing two 18-foot Boston Whalers down the peninsula for sampling purposes. The second Whaler had been picked up earlier that morning and was already being towed south by Danny, the chief engineer of the *Yellowfin*. Danny and Larry were good friends and he had agreed to assist us on the expedition south. This was a huge relief to all of us. Danny could fix just about anything, and taking him along on a road trip into Mexico was like bringing your own pit crew.

We didn't have a lot of time to spend at the boat yard. We were supposed to meet the rest of the caravan down near the border around lunchtime and Larry had already stated he wouldn't wait long. Once we got in, the plan was to load up the Whaler with gear that could safely travel in the boat and then hook it to the tow vehicle and head south. Danny had already called and said he moved the boat into an easy position for the hook up. It shouldn't have taken any longer than five minutes.

When we pulled up to the gate, Tom jumped out of the truck and ran to the front side of the building to unlock the gate and let us in. You wouldn't think that this task would've been life threatening. We heard the screaming first.

From the truck I saw Tom on the other side of the compound running as fast as could backwards. A few seconds later a large pit bull came running around the corner chasing him. The dog's bark was more of a deep, raspy roar and globs of slobber slung from its mouth as it snapped its jaws. There was no doubt that this dog wanted to kill Tom. The dog kept coming and in the fenced off enclosure, Tom didn't have anywhere to go. I jumped out of the truck convinced that this wasn't going to end well. As I got close to the fence, the dog's owner came trotting around the far corner. "LADY STOP!" The big brown dog peeled off immediately and slowly trotted obediently back to her owner. Tom slowed to a trot and finally sat down on the asphalt out of breath. He was close enough to the fence where I could see he was shaking a bit. I felt bad for Tom. We hadn't even left the States yet and he had already had a tough time.

Boat in tow, our group headed south to meet the rest of the caravan. About 6 miles from the Mexican border we pulled into the parking lot of a fast food restaurant. The rest of the group had just arrived and were looking over a map stretched out on the hood of one of the vehicles. It was close to noon and our first stop was still five hours south of the border town of Tijuana. The small town of San Quintin sat on the rugged Pacific shore on the Baja peninsula and was well within the documented range of the spotted sand bass. This was to be our first sample station.

Once we crossed the border and got through the secondary search, we followed the coastal road south of Tijuana. A few miles from the tourist town of Ensenada, the road borders the high cliffs above the rocky shore and gives you an amazing view of the blue Pacific. As we squeezed through the choked town and started traveling the low coastal hills further south, I became excited about the adventure that lay ahead. None of us knew what to expect in the next two weeks of Baja travel. All I can tell you is that we were field biologists headed out to collect samples and I couldn't have been happier.

My travel partner for the trip was my new girlfriend Cheryl. We had first met in an early morning population biology class my first year at Northridge and I instantly didn't like her. She would come into class late after attending a morning dive class with her hair dripping wet. She sat right in front of me and her wet hair would drip onto my desk forming small puddles. I spent a majority of the class flicking the water onto the back of her shirt. Unfortunately the professor teaching the course didn't see the value in this activity and I received a 'C' in the class. Something I still blame Cheryl for.

A year later she showed up as a volunteer on one of the research trips. At the time I found her just a bit too sorority for my tastes and we became nothing more than friends that attended some of the same classes. She was in the same program and I knew I'd see her often. About five weeks before graduating, during a class trip down to a State field station in Mexico, we shed the friendship for something more exclusive. I had no intension of getting involved with anyone during college. I was still healing from a failed marriage, the dissolution the direct result of still being in school at the age of twenty nine and traveling a bit too much out to sea. Cheryl had also recently ended a long term relationship. Being responsible adults at the edge of a new romance, we decided to take it slow. Why anyone starting a new relationship thinks this will work, I do not know.

We pulled into the Old Mill camping area at San Quintin at about sundown. The primitive tent area was desolate and completely covered with ash-like silt, each vehicle trailing a ghostly plume of dust behind it as we made our way towards the harbor. The camp sites were outlined with rocks painted white and provided nothing more than space to set up a tent. An 8-foot-high cinder block wall stretched from one end of the campground to the other, separating the nicer restaurant from viewing the unclean in the camping area.

The caravan pulled up near the harbor. Danny continued on to the ramp to launch the Whaler. Paul, the cook for the trip and another regular from the research voyages, jumped into the truck towing the other boat and followed Danny down near the water. The rest of the group began setting up camp near the cinder block wall.

After dinner, around the communal fire pit, Larry gave us all collecting duties for the following day. We needed to collect 250 specimens of spotted sand bass from each location and each person on the trip was required to participate in the collecting. Danny and Larry would take the Whalers out and see what was biting outside the harbor. The rest of the group would fish the shallow bay from shore. If the fish were here in numbers, we'd bucket them back to camp, set up our work station and began processing the samples.

Towards the end of the evening the group started to settle around the fire pit. We were all beat from a full day of travel, but for some reason we couldn't let the night go. I finished my beer and retired to my tent before the others. Just before I drifted off to sleep, I listened to my friend Ron perform Kool and the Gang's "Celebration" using only his right hand and his armpit.

DAY TWO

June 22ⁿᵈ, 1993

A small gust of wind gently thumping my tent woke me up that second morning. I sat up not feeling particularly great. The dust of the camp was so fine that when the wind picked up, it found its way through the mesh of the tent. Everything inside was covered with a thin layer of dirt and I tasted it in my mouth. I also noticed the taste of tobacco as well.

I vaguely remembered the night before, trying to convince my friend Carrie to quit smoking. She was about to light up when I grabbed her cigarette and tossed it into my mouth. I chewed on the vile stick until it became critical for me either to spit it out or throw up. Carrie had just looked at me, put another cigarette in her mouth and walked off. I definitely got the short end of that stick.

I grabbed my shoes and headed over to the food van. Paul had set up a small folding table and put out a bowl of fruit and several boxes of Granola bars. Next to that was a cooler filled with bottled water. I grabbed one of each and sat in one of the lawn chairs around the smoldering fire and ate alone.

The sun was already up and the primitive camp looked a bit more inviting in the early morning light. Dome tents of various sizes around the camp were lightly dusted and already filthy. All the vehicles looked dull and were completely covered with silt. Just walking to the fire, my pants from the knees down were now dusty and covered in dirt. Anything in camp not covered up was already dirty.

After breakfast I grabbed a fishing rod and headed down to the water. The small harbor was roughly horseshoe shaped, bordered on all sides by boulders. With the number of people we had sampling we could easily fish the entire thing in a day. That was good news if the spotted sand bass were there and hungry.

During my ichthyology class I had learned to target spotted sand bass using hook and line. This species is a voracious, shallow water predator and is easily enticed to bite a lure. If you found them and had enough lures, you could catch them all day long. That is if you found them.

I spent the next few hours fishing in the tiny harbor without a bite. The water wasn't exceptionally clean and the wind had started to push any floating trash to the back of the bay. Plastic cups and gallon jugs floated like tiny icebergs on the mucky surface. Plastic bags and other debris moved below the surface as the tide began to shift. If the spotted sand bass were here, they weren't biting.

On the way back to camp I noticed the Whalers coming back into the mouth of the harbor. Both skiffs were gone when I got up and I figured Danny and Larry had taken them out to fish the outside. As they moved close, I could see that the inside of both Whalers were far wetter than boats should be. The wind guards were covered with water droplets and the decks were shiny and wet. The captains looked a little soaked, disheveled and less than happy as well.

Back at camp things were not looking good. The morning wind had kicked up the dust, making it miserable to be anywhere outside. Most of the group was still in their tents waiting for the wind to die down. The breakfast table had been blown over and every piece of fruit looked filthy and inedible. Literally beat up from their cruise outside the bay, a soaking Danny and Larry reported that no target fish had been caught outside the harbor. My own report was no kind of consolation. The spotties weren't here.

We were slated to stay at San Quintin another day according to our itinerary, but with the weather conditions worsening by the minute and the lack of samples, a unanimous decision to leave was made. Despite the windy conditions, I felt that everyone involved would've persevered and worked up the needed samples. However, if there was nothing for us to collect, there was no reason for us to stay.

Larry stretched out the trip map and checked the route and distance to the next stop. We already knew that we were leaving, so the rest of the group began packing up tents and loading up vehicles. By noon we pulled onto Highway one again and headed south. I never even gave San Quintin a glance in the rear view mirror as we left and I haven't been back since.

Only miles south of San Quintin, the highway begins to pull away from the Pacific coast and starts winding through the low hills through El Rosario. With the vehicles towing the boats setting the pace, the trailing caravan essentially crawled up the first grade towards the interior of the peninsula. I was driving Cheryl's black compact car third from the end and Cheryl was taking a nap in the passenger seat. In the silence of the drive, San Quintin was still in my mind. Our first stop on this trip had produced no fish samples. When our graduate group began planning the expedition, we did plan for the possibility of areas not producing. But having the first spot come up blank left me feeling a bit discouraged.

About 240 miles south of San Quintin was the town of Guerrero Negro, the next stop for our caravan. After winding through the foothills into the center of Baja and traveling through the small towns of Punta Prieta and Rosarito, Highway one arcs back towards the Pacific Ocean. Guerrero Negro sits on the coast, at the edge of Laguna Ojo de Liebre, also known as Scammon's lagoon. At one time the coastal town was the center for most of the salt processing along the peninsula. The abandoned salt plant was where we were headed. We had received information from the Mexican University located near Ensenada that spotted sand bass were regularly caught in the shallows near the old salt processing plant. When Larry asked if our group would be able to camp near the abandoned building, the Mexican student just laughed.

The scenery of the interior of Baja was amazing. Ruby red rock formations dotted the desert, completely surrounded with large saguaro cactus, some over thirty feet high. Like giggling school children we'd occasionally pull over to quickly snap a photo of one that looked like it was giving us the finger, or others that were shaped like a huge penis. The clouds formed by the gulf would stall at the coast and made every scenic turnout as pretty as a postcard. It was almost a shame that we had to drive the road without stopping for awhile to enjoy it.

The route through the center of the peninsula was not without casualties. Shortly after leaving the coast, I turned a corner and surprised a small covey of California quail slowly crossing the road. I was on the birds too quickly to stop and I plowed right through all of them leaving a cloud of feathers and bird parts behind me. One of the bird's feet and part of a leg became stuck on the wind shield wiper. Just before the collision I was riding high on adventure. I was excited to be a new biologist in a new land traveling the back roads of Mexico for science. Bashing through a covey of birds made me realize I needed to be careful down here. We were in the middle of nowhere and even slight car trouble could be disastrous. Despite my renewed determination to stay safe, I continued to plow through the wild things of Baja.

Twenty minutes down the road I hit the largest cottontail rabbit I had ever seen.

Smaller animals, with smaller brains, also marked our passing. Kangaroo rats dotted the road's edge and scooted across the highway when the caravan neared. For some reason, instead of racing completely across the road, they'd stop in the middle and head back to where they came from. Not many made it back. There were too many to maneuver around and the sound of popping rodents became a regular noise echoing from the undercarriage of the car.

The road out of the foothills and back towards the Pacific coast was long and straight. The row of cars, trucks and boats coasted the two lane highway down towards Guerrero Negro and seemed to follow the sun. We were definitely blazing a trail. None of us had ever been down this road before and with the exception of San Quintin, all the locations we had chosen for sampling were selected sight unseen and by map. We had no information on accommodations or if the land was private or public. Being nomadic and in the spirit of adventure, we had to be flexible with what we found when we got there. And I really don't think any of us cared. With each stop for food or gas, the group would gather in an excited bunch. Each of us was thrilled to be on the trip and we couldn't wait to see what awaited us at the next stop.

We were still about 100 miles from the town of Guerrero Negro when we pulled over for a snack at a turnout along the straight road. Just before we had left San Quintin, Larry had given Mike Franklin a call and let him know our change of plans. Family commitments had kept him from starting the journey with the rest of the group and now he was in route to meet us. Mike and I had met at the university and pretty much became friends instantly. When I was trying to figure a way out of my old engineering life, I had called the university to hopefully get some guidance. After ringing just once, Mike, a graduate student just finishing up his research, answered the phone. I credit Mike with putting me in contact with the right people and guiding me through the acceptance process. I can say emphatically, if it wasn't for meeting Mike, I wouldn't be where I am today.

Mike wasn't on the road alone. He'd be traveling with our hired gun for this research trip. Mark was an avid spear fisherman and was hired to shoot anything that could be of interest to science during our Mexico adventure. Since I also enjoyed the sport of spear fishing, I was looking forward to meeting Mark and maybe picking up some pointers.

Just an hour before sundown we pulled into the town of Guerrero Negro. The little city was alive with activity. Small shops and bakeries were busy selling whatever they had left at the end of the day. Dogs of every color and size trotted the sidewalks or found slivers of sunlight on the bare ground to lie down in. Small children in dirty clothes swarmed around the caravan as we moved through the main street, begging for whatever wasn't tied down.

We pulled to a stop in a large open lot next to a few shops. Since the food wagon was all packed up, we were on our own for dinner. Larry instructed us all that we should grab something to eat and then meet back at the cars shortly. Greg, John and I walked across the street to a colorfully painted taco stand. A huge painting of a manta ray jumping from the blue sea was displayed on the side of the little food shop. Above the painting were the words, 'Manta Ray's'. The taco stand was located in a vacant lot, constructed of little more than four pieces of plywood. It had a dirt floor and the flies seemed to really like the place.

The three of us sat on the wooden bench and ordered a few fish tacos each. Through the course of the meal, we found out that the owner and cook of the small shop was indeed named Ray. The tacos tasted good and were cooked perfectly. While we ate, we tried to figure out what type of fish we were eating. The word 'fish' has a very loose interpretation when eating anywhere in Mexico. There are only two things you can be absolutely sure of when a plate of 'fish' is placed in front of you down in Baja. It came from the sea, and it used to be swimming.

To John and me the type of fish used to make the tacos didn't matter. They tasted good and that was all we needed. Greg seemed a bit more concerned about the ingredients in his food. Greg spoke Spanish better than the rest of us and asked Ray about the fish in the taco. The cook looked up from his huge vat of oil smiling and walked over to the side of his little taco stand. He then pointed at the huge picture of the manta ray painted on the plywood. "Tacos mantaraya," he stated. We had just eaten manta ray tacos. Welcome to Mexico!

After dinner we loaded back into the cars and headed out towards Scammon's lagoon. After a few wrong turns and some confusion, we finally found the abandoned processing plant a few miles outside of town. The structure was covered with graffiti and every window had been broken out. The main plant sat right on the lagoon and at one time had been the center of all major salt processing in Guerrero Negro. Now abandoned, the building sat decaying at the edge of the sea.

Down by the shore four large cement platforms, about forty feet in diameter and twenty feet above the water, faced the open sea and lined the beach in front of the processing plant. These structures were covered with rusted plates of metal and large rubber bumpers to protect the ships that used to dock there. I walked out to the edge of one of the huge platforms and took a look around. The abandoned towers would make great fishing platforms if the target species were here.

Back at camp, everyone was busy setting up tents. Paul had prepared some snacks and Danny already had a nice fire going in a makeshift pit. Lawn chairs already surrounded the fire and Larry was getting fishing gear ready for sampling.

It got dark before we could head down to the shore and test the fishing. Larry didn't want any of us stumbling around in unfamiliar surroundings trying to fish, especially around the large cement structures near the beach. After camp was set up, most settled around the fire. It had been a long day of travel and I was relieved that my only required task for the evening was to relax. While I watched the fire burn down, I found myself getting anxious. I hoped that the spotted sand bass were here.

DAY THREE

June 23rd, 1993

Noise from the shore woke me just as the sun was peaking over the horizon. I could hear muffled yells in the distance, but I couldn't quite make it out. It sounded like Larry and Danny were laughing and yelling at each other from down by the beach. Just outside of my tent someone had placed a five gallon bucket filled to the top with writhing fish. They were all spotted sand bass.

I grabbed my gear and followed the racket. Larry, Danny and Jimmy were standing at the edge of one of the cement pillars fishing. Two of the fishermen were hooked up. A row of five gallon buckets were lined up behind the trio and one was already full of spotties. I took a spot on the end and made a cast. My lure never hit the bottom. An aggressive fish grabbed the lure and I set the hook. Less than a minute later my first spotted sand bass of the trip was splashing in the bucket with the rest.

For the next several hours we caught spotted sand bass almost at will. All five buckets contained fish and most of them were just about full. By mid morning nine fishermen were spread out over the platform assisting in the sampling. Surrounding the buckets there were hundreds of lure pieces, a few floating in with the fish. I leaned over the closest bucket and made a rough count of the samples. With the bucket back at camp, I knew we were real close to the number of fish we needed.

Larry caught me looking over the fish and suggested we start processing the samples. I knew the fun of collecting specimens was over. With the help of Greg we carried the four full buckets back to camp to begin the work-up.

We set up an assembly line to work up the fish. I was in charge of weighing and measuring each specimen and removing the gonads. Greg would extract the valuable aging bones with a set of forceps and place them in a numbered coin envelope. Carrie was set up right next to me and would assist in bagging the larger tissue samples in a chemical preservative. John set up a lawn chair right in front of the work table and would be in charge of recording the data. Ron and Andy volunteered to fillet each of the larger specimens once science was finished with them. The rest of the group was tasked with keeping the fish coming and making sure everything else was in order. After we set up the individual sample stations, it was time to get started.

I grabbed the first fish and stretched it out across the measuring board. The mouth of the dead spotted sand bass was agape. I read off the length to John and then transferred the fish to the digital scale. After I read off the weight, I moved the fish back to the measuring board and pushed a small pair of dissecting scissors into the vent and cut the belly open all the way to the throat of the fish. I pulled out the two gonad lobes and snipped the tissue holding them in place. Carrie held open a sample bag filled with preservative and I dropped both into the bag. I then scored the auditory sack just behind the head with a knife and cracked it open over the edge of the cutting board. I tilted the fish towards Greg and he gently pulled the otoliths from the head with a pair of tweezers. He dropped the aging stones into a small coin envelope and we were done. I tossed the fish into a fillet bucket and grabbed another. On this day, I would repeat this procedure 265 more times.

In the monotony of processing fish samples time loses all meaning. The day's events were slipped into the dissecting machine and we scarcely noticed. About half way through the second bucket of fish, Paul served us a quick lunch of cheese and crackers. If you're going to study fisheries sciences, you better get use to not only eating with dirty hands, but eating in close proximity to some pretty nasty stuff. At the end of the third bucket, Eric, Larry's thirteen year old son, told us he was headed down to the lagoon to snorkel. And as we were working on the last few fish, Larry showed up and said that the sampling for science was over and what we had in camp was all we had to process.

We had been working non-stop for hours and the mundane, stinky jobs started to take their toll on us all. My back ached from leaning over the measuring board and my shorts were covered with fish slime after frequent contact with the soggy work table. Carrie was now chain smoking and the cigarette smoke was wafting right over the work area. Greg had been silent for over an hour and you could feel that the monotony of the tasks was getting to everyone.

To speed up the whole data reporting process, I would yell out the data for each specimen and rely on John to write it down. Every so often he would repeat a questionable length or weight and I knew he was paying attention. For the last twenty minutes John had been silent as well. Apparently he was not immune to the tedium of the tasks either. The group was on the last bucket and by my estimation we had another twenty fish left and we'd be finished. I reported the specimen number and happened to glance over to John. His lawn chair was tipped back and he was balanced on the back legs. The data book was in his lap and his pencil sat ready for the next data point. Unfortunately, the pencil operator needs to be awake for data to be accurately recorded. John had fallen asleep.

No one had bothered to make sure that the data was being recorded and the information for the last five fish samples was missing due to John's nap. This was not the only data inconsistency. The current specimen number on the otolith envelope was also off. We were tired and now we were making mistakes. Each of us had been doing the same stinky job for hours without a break and it started to show.

Cheryl took over for John and we pushed forward to finish the last few samples. The fish with the missing data were removed from scientific consideration. The otolith envelopes were renumbered to match our current position and we continued on. It would be two full years before I would find out why the envelope numbers did not match. As I worked up my specimens back at the lab, I opened up one envelope and poured out three pairs of otoliths, instead of the standard one pair. Apparently John wasn't the only one snoring on the job. Instead of switching out otolith envelopes after each fish, Greg, hypnotized by the routine, had deposited three sets into the same envelope.

The last fish, fish number 265, was run through the sampling machine and the carcass was tossed lifeless into the fillet bucket. It had taken us eight hours total to catch and completely process all the samples. Six and a half hours of that had been spent working up the fish and recording the data. To this day I have never worked through as many fish in a single session as we did that day in Guerrero Negro.

I stowed the data book in one of the wooden sample boxes and grabbed four cold beers from the large cooler in the back of the food van. I walked over to the edge of camp to see how the final phase of the process was going. Andy and Ron were over near the lagoon finishing up the filleting. A white bucket sat between the two filled with water and at least fifteen pounds of spotted sand bass fillets. The fronts of their shirts were smeared with dried fish blood and entrails. I cracked the caps on two of the beers and placed them on the fillet table. With slimy hands, both gladly took a beer break.

Our camp and where we had set up our work area was in the shadow of the old salt plant, in what I can only imagine used to be the parking lot. Ron and Andy were at the edge of camp cutting up the spotties on an old cement structure that used to be a large cooking pit. The pit sat on a bluff that overlooked a smaller lagoon behind the large cement structures. Pangas and other smaller boats were anchored in the crystal clear water of the little lagoon. Boats weren't the only thing floating in the bay. Drifting around in the shallow water was thirteen-year old Eric. Thinking back to when he told us he was going snorkeling, I figured Eric had been in the water for over four hours. And from the looks of it, every single second of that he had been without a shirt.

In Baja you quickly realize that in the lower coastal areas shade is a rarity. Most of the vegetation is less than three feet high and the taller saguaro cactus offer enough cover from the sun for only one person, and he needs to be standing. You begin to see why most of the natives stay inside during the mid day heat and take a siesta. Some of us didn't know any better. Eric had taken a dip in the cool bay waters to get relief from the mid day heat. I think most of us would've done the same if we hadn't been tied to the work tables. Although I think I would've at least put on some sunscreen. From where I was standing, Eric looked like he was swimming around in a pink shirt.

That evening, to celebrate our first successful sampling station, we headed into town and ate dinner at one of the local restaurants. The burritos and tacos were fried in lard in a huge vat that sat within feet of the dining area. The two items, along with rice and beans was all the small café served. The food was amazing. We drank cola from banged up bottles that had seen a thousand set of lips before ours. We ate and drank whatever they brought to the table and essentially shut the place down. Guerrero Negro had shared her bounty with us earlier that day, and now we wanted to return the favor.

There were two members missing from our party that night. Jo, a volunteer for the trip and a relatively new face to the fisheries program was unhappy with the accommodations of the trip and the quality of the food. She had started to voice her concerns back at San Quintin, but not many had listened to her. As our Mexico trip unfolded, most started to realize that the only outdoors that Jo had ever experienced was when she traveled between malls. She had little camping experience and roughing it in Mexico was not the time or place to figure out you didn't like it. She had decided to stay back at camp while we celebrated. Eric was also back at camp. The sun had set on his afternoon swim and his young body began to pay for his neglect. Feeling the effects of sun poisoning, he was sacked out in his tent to ride out the basting.

When we returned to camp we settled around the fire and let another day slip away. We gathered in a circle around the pit and shared stories and spirits. This trip was the first time that our graduate group had traveled together for any length of time. Free from the strict university policy and in another country, a few began to spread their wings. Greg is a soft spoken individual, quick of wit and as bright as they come. One thing Greg is not is a drinker. Suckered in by the rich, sweet taste of rum, Greg began drinking. His sips turned to gulps, and as alcohol invaded his body, likely for the first time, he began to experiment. Towards the end of the evening he was convinced that dunking an Oreo cookie in with his rum was one of the best combinations ever.

Sometime during the evening the wind shifted and the gentle and ever present stench of urine wafted through the camp. By now most were used to the changing aromas of Mexico and with our day's success, there wasn't much that could dampen our mood. This collecting site had gone perfectly and we were now free to enjoy Guerrero Negro. The group enjoyed each other's company around the fire that night and speculated on our next stop on the peninsula.

DAY FOUR

June 24th, 1993

The next morning broke far quieter than the previous. A few of the group rose with the sun and took one of the Whalers out to sample some of the offshore fishing. Tom had gotten up shortly after to dive the fishing platforms. The rest of the group was naturally slow to rise.

On my way to the food table, I noticed Greg on his hands and knees digging around the outside of his tent. I walked over and saw that he was using a stick to cover up several rather large piles of gray vomit. He looked over at me and I could tell that he still wasn't feeling great.

After I helped Greg bury the evidence, I walked down by the beach. The tide had shifted to low and areas that were once feet under water were now exposed and accessible. I carefully hopped my way out onto the rocks and explored the tidal pools. The sun peaked over the bluff near camp and warmed the air around me almost instantly. I could feel it was going to be another very hot day.

I spotted Tom further out floating at the edge of one of the platforms. I was close enough to hear his breathing through his snorkel and I could tell the way he was hyperventilating he was getting ready to dive. Within seconds he arched his back and quietly disappeared under the water. I decided to join him.

I rummaged through the back of Cheryl's car collecting my dive gear. I told John where I was headed and he decided to join me. Next to the car and a few feet from where I was parked was Eric's small dome tent. Earlier that morning I had seen the Whalers head out for a day of fishing and was almost certain Eric had gone with his dad. John mentioned that he must've been feeling better. After a few minutes, I noticed some movement inside the tent. Shortly after that Eric stumbled out. He was clearly not feeling better. The skin we could see was bright pink and he looked to be sweating. Both of his hands cradled his belly and he just stood there looking at the ground.

Eric was a lot younger than the rest of the group and seeing that his current guardian was out fishing, we were pretty much responsible for him. I walked over and asked him how he was doing. He didn't answer me and just kept staring at the ground and kind of breathing funny. Carrie walked over and gently rubbed Eric's back. For some reason our concern annoyed him and he started demanding something to drink. I figured at thirteen he was too old to throw a tantrum, but he sure was giving it his best shot. It was clear to me that he was angry for some reason and I didn't feel like babysitting him.

Carrie brought him a large glass of milk. Eric roughly grabbed the glass from her and began to take huge gulps, spilling most of it on his face and shirt. I was done. I walked back over to the car and grabbed all my dive gear.

Eric finished the milk and demanded to know where his dad was. Carrie told him that he had gone out fishing in one of the boats and that he probably wouldn't be back for a while. This made Eric angrier. I was more than willing to take care of Eric while his dad was out, but I was not going to stand there and be ordered around by a kid. John and I disengaged ourselves from the situation. I didn't want to deal with Eric's attitude and John absolutely did not care. As it turned out, the situation would kind of take care of itself.

Eric shuffled over to the group sitting around the morning fire eating breakfast. The sun was warming everyone up and further toasting the now exposed Eric. Most weren't paying much attention to him. That is until he started yelling.

Eric stood at the edge of the group and began screaming something that I couldn't understand. He closed his eyes tightly and yelled again. His volume and slurred speech kept us from comprehending his demands. We all looked at each other and absolutely no one moved. "What did he say?" I muttered. John once again shrugged. Sensing that no one was getting or caring what he was saying, Eric stated his demand again. This time he clenched his fists and stomped in place. "IM-GONNA-THROW-UP" he screamed. At that point I think some of us got it. Carrie walked up to him and rubbed his back gently, trying to calm him down. Eric instantly turned towards her and promptly vomited almost half a quart of warm belly milk all over Carrie's shoes.

Carrie cleaned Eric up and helped him lay back down in his tent. She told us she'd be in camp for the day to look after him. Eric had clearly received a serious sunburn from the day before. But I started to think that maybe Eric had also suffered sun stroke during his swim. I had suffered sunstroke when I was fifteen while on a hunt with my uncle. I remember throwing up several times and shivering for almost two days. They kept me hydrated and I got lots of rest. But that was back in the states. We were worlds away from any quality health care or help here in Mexico. Even though Eric had become belligerent, we all knew it was because he wasn't feeling well and we'd definitely needed to keep an eye on his condition.

John and I walked down to the beach and got suited up for the dive at the edge of the lagoon. Tom was just getting out and walked over. Despite his empty fish stringer, he said that he saw lots of fish and that the visibility was excellent. He pointed out some areas that may be of interest to us, warned us briefly about a developing current and then started walking back to camp.

We waded into the lagoon formed behind the last large cement cylinder in the line of four. Almost immediately we were met head on with an extremely strong current. Small fish and debris swept passed us as we struggled to get around to the front side of the huge round structure. The current was blasting into the right side of the huge cylinder and splitting its velocity between the front and the back of the platform. A smarter diver may have turned back here. But me and my Korean counterpart were neither afraid of a little current nor considered smart divers.

I pulled myself along the rocky edge trying to move forward. The current was too strong to swim against and I didn't feel like turning back. We had been swimming against the current funneled towards the back of the cylinder for 10 minutes and making very slow progress. At times we had to actually pull ourselves forward along the back of the cement cylinder. After a constant struggle, I finally reached the apex of the cylinder and was instantly pinned to the rusty structure by the force of the moving water. Luckily, I was pinned at the surface, but I couldn't move and there was nothing on the smooth cement surface to grab onto.

Using my thick diving gloves, I grabbed onto small barnacles and slowly inched myself forward along the rusty structure. The strong current thumped my head and mask against the cylinder regularly. My snorkel was ripped from my mouth and now bounced rhythmically against my chin. My black wetsuit was now covered with large brown patches of rust and I was completely worn out. The dive was no longer about exploring the bay. We needed to get back to shore and safety.

After a few more minutes of pulling myself along the rusty tube in the turbulent water, I felt the bulk of the current finally propelling me down the front side of the structure but a lot faster than I wanted. I knew that I would probably be deposited into the eddy that formed between the platforms and I was hopeful that the whole ordeal would be over soon.

I sped past the front of the structure and was instantly halted and slammed into the front of the fishing platform. Even though my back took the brunt of the impact, my head connected with enough of the cylinder to cause me to see stars.

The towline on my spear gun had apparently caught on something on the other side of the death tube and I battled to maintain my grip on the slippery weapon. The tremendous current was unrelenting and being dangled at the end of a line in the white water was not good. As I desperately tried to free the rope from the obstruction, I kept bouncing violently into the cement tank. I had both hands on the gun and I tugged with all my strength to pull the rope free. If there is one thing I can say for certain, it was that now I had absolutely no thought or concern for John or where he was.

I gave a few more stern tugs on the line, but it wouldn't budge. Just as I was attempting to leverage myself against the platform to snap the rope, John made an appearance. He raced by me at a high velocity, free of any lines. He appeared from the gloom tumbling ass over elbows. Our eyes met briefly as he passed and he looked to be smiling, and then he was gone.

I knew I had to free the tag line or drop the gun. My mask had sustained a direct hit from the platform and my right eye was now swimming in seawater. I reached up and tried to straighten my mask and snorkel and almost lost my grip on the spear gun. That was enough. I reached down, grabbed my dive knife attached to my calf and cut the line that held me. Once the line snapped, I caught the full power of the current in the chest and I was now completely helpless. I put all my faith in the power of the water and hoped against hope that I didn't slam into the pier or John's floating body. While the current raced through the two platforms, I felt the velocity ease a bit and noticed that I could feel the sandy bottom with my fins. I was still being pushed, but with a few kicks I found myself in the still water between the twin pillars of death.

I pulled myself up on some rocks. I tossed my spear gun up on shore and pulled myself completely out of the moving water. John was perched on a small boulder a short distance away laughing hysterically. I glanced down and saw that I was completely covered from head to toe with rust smudges. My mask was crooked on my face and my snorkel sported a nice 90-degree bend in it. I had been beaten up pretty badly and I looked like I had been dragged behind a truck in the mud. I looked at John through my one good eye and started laughing myself.

I grabbed my gear and we limped on back to camp. I was trying to fabricate a story that sounded more heroic than what I had just been through, but my appearance was the only explanation needed. As we turned the corner to the camp, everybody started laughing.

John and I cleaned up and got something to eat. I spent the rest of the day fishing for pleasure and enjoying the tempo of Mexico. John and a small group decided to head down to the mud flats and dig for clams. Since we had collected all we needed from Guerrero Negro the first day, today and tomorrow were slated as fun days. And while I really enjoyed all activities on the ocean, I did think that I would take a break from diving for a while.

Shortly before dinner, John and Carrie returned with a bag full of heart clams and a few large crabs that didn't make it out with the tide. John placed the overloaded bag on the cooking table. I walked over to the writhing bag and took a peek. The sand-covered clams and crabs looked less than appetizing, and I wasn't quite sure the clams would be safe to eat at this time of year.

During some summer months, red tides comprising of billions of small marine organisms termed microscopic algae can flourish in the shallows in warmer climates. Each little animal contains a toxin that can be concentrated in the flesh of filter feeders like clams when they consume the algae. If humans consume the clams, the concentrated toxin can be harmful, if not fatal, and result in a condition known as paralytic shellfish poisoning. I had read enough about the condition and its effects to know I had no desire to risk it. I also didn't feel like going home without my legs, since even a mild poisoning usually resulted in the amputation of the affected persons limbs to save their life.

Just the suggestion regarding the clams in front of a handful of marine biology graduate students planted enough of a seed of doubt for us all to gather around the cooking table. After several minutes of discussion, it was clear that none of us had any idea whether the clams would be safe or not. We decided to err on the side of caution and the entire day's haul of invertebrates was deposited into the blue sample cooler along with several bags of spotted bass fillets.

A few hours after dinner I noticed a lone set of headlights breach the top of the hill coming from town. Mike's silver Toyota two-door pulled up and slid to a stop several feet from the fire. Mike exited the vehicle, leaving the headlights on and the door open. He walked over to the fire pit and sat himself down in a vacant lawn chair next to me, fatigue and the wariness of 600 miles etched on his face.

"You sir, look like you could use a beer," I said.

He let out a deep sigh and looked my way. "That would be a start."

I handed Mike an open beer. He grabbed it, not really concerned with the brand name and chugged the entire thing right in front of me. As he finished, he wiped his mouth on his sleeve and without looking, handed me the empty.

If you didn't know the travel arrangements, you would think that Mike had made the lengthy road trip alone. His car, lights on and door open remained dark and silent for several minutes. That is, until Mark opened up the passenger side door in the dark and exited the vehicle. The two had driven non-stop from Los Angeles to Guerrero Negro in a little less than 16 hours.

Mark sheepishly exited the car and stood next to the dusty hood. The slamming of the passenger door caught Mike's attention and he reluctantly lifted himself from the chair to give the introduction. With minimum enthusiasm, Mike motioned to the figure near the car and introduced him to the group. With that, Mark entered the lawn chair fire circle. Pleasantries were passed around. Mike asked me to get him another beer.

I could feel the day's activity pushing me towards sleep. When I got up to head to the tent, I felt a twinge of pain in my shoulder, an obvious product of the day's diving adventure.

Tom, not wanting to miss the opportunity to once again rub it in, spoke in a slightly beer-soaked voice. "Maybe you should leave the diving to me, Timmy," he said, smirking and nudging whoever was next to him.

Tom was an experienced diver and if I had been smart, I would've heeded his advice on the current. I lifted myself from the chair and nodded to Tom. I conceded to his good natured ribbing and headed to my tent. I think the reason Tom and I got along so well is that we both enjoyed joking around with each other. I expected nothing less from him. I also knew for the remainder of our stay in Guerrero Negro, instead of stumbling to the distant dunes in the middle of the night to relieve myself, I'd be peeing right next to Tom's tent.

DAY FIVE

June 25th, 1993

The next morning I was extremely sore from the diving activities of the day before. My back and neck ached and both my biceps burned. My right elbow had a knot on it and extending it out completely took some painful effort. I ached just about everywhere. I slowly exited my tent and saw that Mike was sitting in the fire circle of the night before. I walked over and eased my tired body into a chair. I found a comfortable position and turned my face towards the rising sun. Mike had watched my tortured trek from the tent to the fire.

"What the hell happened to you?" he asked.

I decided to skip the truth.

I caught Mike up on the progress of the trip and how we'd done so far. We sat by the dying fire munching on the ever-present granola bars and talking. The camp was deserted except for the two of us. After some convincing, I got Mike to tell me what had happened on the trip down with Mark. He rubbed his face with his hands and stared off towards the bay. He then spent the next fifteen minutes explaining his driving journey down from the states. It wasn't pretty.

Mike explained to me that Mark is a music composer by trade and enjoys the sport of free diving to relax. Apparently Mark hadn't been in the car for five minutes when he explained to Mike that due to his extensive musical background, he had become what he called music satiated. Meaning, he couldn't listen to music. No radio, no singing, nothing. "Have you ever tried sitting in a car for 16 hours without listening to the radio?" Mike asked.

After breakfast Mike and I headed down to the shore to fish for fun. We spent the next few hours casting and no less competing. It seemed that none of us could pick up a rod without someone else taking it as a challenge. When it got closed to noon, we called it a draw and headed back to camp for lunch.

On our way we saw Mark returning from his dive around the cement towers. The first thing I noticed was that his wet suit didn't have a rust smudge anywhere on it. The second thing was that his stringer was empty. Mike told me that Mark is very experienced at spear fishing and only shoots very large specimens. I figured that he didn't encounter anything large enough to shoot, or that he was also fish satiated.

I walked over to talk to Mark. I was curious to see how the dive had gone. After I asked, Mark dropped all his gear and stared out over the lagoon. He paused, scratched his chin and sighed deeply. As he searched for the words to describe his experience, I realized that I had already lost all interest in my own question.

On the way back to camp we talked about diving and spear fishing. During the short walk Mark described an encounter with a 100-pound grouper that hid in its lair deep beneath the fishing platforms. Mark said he sat on the bottom and just watched the large fish in its cave. I knew Mark was on this trip for one reason; to shoot fish. And if Larry heard that he had passed up the opportunity to land a large grouper, a fish that could feed our entire group for a week, satiation would be the least of Mark's problems.

During lunch Larry told the group collectively that we weren't quite done with the fish sampling for Guerrero Negro. Part of Greg's research not only focused on comparing the physical morphometrics of the different spotted sand bass populations, but he also wanted to look at the genetic differences as well. This was going to require blood samples from live fish from each location. To avoid contamination during the extensive fish work up, Greg had opted to wait until we were done to collect what he needed. Since he was looking for a general genetic profile of the population he didn't need an extensive amount of blood samples. After some discussion, Larry decided that blood from twenty different fish should be sufficient for Greg's research.

After I was finished eating, I volunteered to catch what was needed and Mike quickly joined me. Greg handed me a sample collecting kit and I met Mike over near the school tackle to grab some fishing supplies. We each grabbed a bag of lures, not really concerned about the style or color, and headed out to fish for science. It was our last midday at the lagoon and I was glad to be spending it with Mike.

On our way out we heard a heated argument boil over around the fire pit. Larry was standing over a seated Mark, red faced and angry. We caught enough of what was being said to know that Mark had just told Larry about the big grouper that still lived under the platforms. From the posturing, you could tell that Larry was explaining to Mark in angry detail exactly what his role was on the trip.

By the time Mike and I returned from sampling it was close to dusk. It hadn't taken us long to catch all the fish we needed for Greg's research and we spent the remainder of the time fishing for fun. I took the sample fish and removed blood from each by simply snipping a gill arch and dropping the tissue into a vial of buffer. I loaded the sealed vials into a plastic bag and we were done. The fish usually survived this procedure and everything we caught was released.

I placed the bag of blood samples in the blue sample cooler. This particular cooler contained a huge block of dry ice and was to be used exclusively for the research samples collected on the trip. When I opened the lid, the collage of colors and smells that radiated from the small icebox had me taking a few steps back. John's clams and crabs from the day before were now stewing in their own grayish-brown ooze. A small bag of fish fillets had opened up and several of the fleshy strips were floating like little islands in the rancid fluid. Wedged down between one of the last pieces of block ice was a silvery, foot-long fish with its guts hanging out. A fish shot by Tom that may have been a range extension for the species. Since there were food items in with the samples, I wondered if some weren't clear on the role of the blue cooler. I placed the bag of samples on a piece of block ice and closed the lid. The contents looked like a simmering pot of primordial stew.

I cleaned up and met everyone around the evening fire. This had been a good station, but we were done here. I was anxious to get to our next spot. Ron and Andy had employment obligations back in the States and were busy packing things up for their return trip the next morning. While I was helping them load their gear into the truck, I noticed Mark walking back from a dive. He had wisely decided to head back out into the bay to try his luck again after his angry interaction with Larry.

Skirting the fire pit where Larry was sitting, Mark walked up to the back of Andy's truck. Hanging from his stringer was a six pound leopard grouper. Mark's extensive experience as a spear fisherman was being severely hampered by his inability to identify fish species that may be of interest to our expedition. The leopard grouper was not only common for the area, but we had landed over a dozen during the wide open spotted sand bass bite a few days earlier. The fish was in perfect condition and I didn't see any spear hole. Mark explained that he always aims for the eyes when hunting underwater. Sure enough, the only mark on the grouper was a spear hole through both eyes.

I didn't have the heart to tell Mark that the grouper was common for the area and absolutely no use to us except for fillets. He seemed like a sensitive guy, and being tossed into our group as the only stranger, I guess I felt a little sorry for him. He unclipped the fish from his stringer and held it up. I knew the fish should be taken straight to the dinner table to be carved up for fish burritos, but I told him to put in the blue cooler instead.

We spent the evening taking pictures and drinking beer. A few of us toasted Ron and Andy and tried in vain to persuade them to blow off their jobs and stay with us. By midnight Larry convinced us all that we were slated for an early start the next morning and the evening celebration was over. The group disbanded to their dusty tents.

Through the flimsy vinyl I watched the stars shine on Guerrero Negro. As the fire popped, I thought about life. I thought about how tough it had been to take the things in my life that I knew were wrong and change them. At both a personal and professional level, they had been the hardest things I had ever had to do. But those changes had to be made if I was going to live the life I wanted and be happy. I was confident that this new path I had chosen was the right one for me. Despite lying in a dirty tent in Mexico, downwind of a stinky building, I couldn't have been happier.

DAY SIX

June 26th, 1993

The next morning was full of activity. We ate breakfast as we packed and tried to leave the Guerrero Negro salt plant as clean as we found it, which wasn't going to be difficult. Larry and Danny had the map out and were carefully plotting our next destination. After an hour of gathering trash and making sure everything was secure, we were ready to head back towards the small town of Guerrero Negro and Highway One.

Back in town we grabbed a few supplies, gassed up the vehicles and slowly moved through the little Mexican town. Just outside of Guerrero Negro, Larry pulled the group over and made sure we all knew where we were going. Jo's car was parked behind Cheryl's and I noticed that she didn't get out of the car during the road stop. Greg was riding with her and said she wasn't feeling good. I think at that point most of us were tired of her complaining. From the beginning of the trip she had found several things that didn't meet with her standards. Unfortunately, both Greg and John were riding in her car and we couldn't just send her home. It was also made clear at the beginning of the expedition that we all stay together. Sending a young woman back to the states, a two day drive, by herself was out of the question. She was stuck with us and we were stuck with her.

Our next stop was the Bay of Conception on the gulf side of the peninsula. The small town of Punta Chivato is located at the northern portion of Bahia De Santa Ines, right at the mouth of Bahia Conception and near the town of Mulege. This would be our first stop on the gulf side of Baja and our first gulf station.

As soon as you leave Guerrero Negro, the highway leaves the Pacific side of the peninsula and heads through the cactus desert and passes through the town of San Ignacio. Continuing east, the two-lane road begins to border the bright blue Gulf of California near the town of Santa Rosalia. That's where we got our first glimpse of the Sea of Cortez. We fueled up there and after an hour of winding through the coastal hills, we reached the dirt turnoff towards the bay at noon. Another hour's travel down a poorly graded road and we rounded the corner to Punta Chivato.

There were several sailboats and a few pangas anchored close to shore in the protected cove. We could also see two large submerged reefs in the crystal clear water a short distance off shore. Darting around the rocky structures were several fluid shadows that could only be fish. Some of the shapes were larger and seemed to be swimming alone.

The little bay was flat calm and the clear water was more than inviting. It was at least twenty degrees warmer on this side of the peninsula and exploring the shallow lagoon after we set up camp sounded like a great idea. I have no idea why I was even considering a dive after my dance with the tubes of death at Guerrero Negro. The calm gulf waters did looked far more inviting than the frigid pacific had when it had tried to kill me.

Down at the beach we made arrangements with the custodian of the camping area. For a small fee he allowed us to set up camp in a dirt lot near a small grove of palm trees. Next to the lanky trees was a small cinder block building that contained working showers and functioning rest rooms, a luxury that had not been available on the trip until now.

The group set up camp in the mid day heat. The refreshing warmth of the gulf that had greeted us at the top of the hill was now stifling. There wasn't a stitch of wind and the only way to cool off was to get into the water.

Once the camp duties were finished, Cheryl, John, Greg and I grabbed our dive gear and headed down to the nearby lagoon to cool off. The warm gulf waters meant we could leave the heavy wet suits behind. The short piece of frayed line attached to my spear gun was a reminder of my last dive. The gun itself was covered with rusty smudges and several new scratches. I had been on tough dives before, but that last one had banged me up pretty good. I knew that with more experience, I would begin to identify questionable diving conditions and just wait for another time. Or pay more attention to experience divers when they gave me advice.

Larry made sure we knew we were still on the clock by strongly suggesting we bring something back for dinner. Of the four of us, John and I were the only ones armed, so the dinner duties fell on us. We had already eaten leopard grouper and spotted sand bass on the trip, but I was hoping that the gulf waters would provide us with something larger and more exotic.

Greg and Cheryl stayed in the shallows. John and I headed for the deeper reef out beyond the anchored boats. The water was almost too warm at about seventy five degrees. The visibility in the lagoon was close to 100-feet and large schools of fish were everywhere. Within minutes of leaving the shore, the dark brown rock of the deeper reef came into view. When we got close, John and I split up and began to search for fish. It was time to hunt.

The reef stretched from the bottom twenty feet below, up to within a foot of the surface. Large buoys and thick buoy lines marked the rock at the middle and at both ends so boats avoided collisions. Dozens of species of fish darted in and around the reef, defending their tiny territories. The barred sergeant majors moved in loose schools protecting a few larger rocks towards the bottom. Several species of wrasse swam their way through holes in the rocks, chasing and being chased by different fish. Large trigger fish hovered and then chased other species of fish away from a territory that appeared to have no definite boundaries. Occasionally a larger shadow would dance off the bottom near the edge of the reef; leopard and broomtail grouper waiting for prey to venture too far from the safety of the rock. Life and death was everywhere.

It's hard to explain the feeling I get when I enter the water. Sometimes I think I don't fully understand it myself. I enjoy testing myself on the identity of the species I encounter. At the very least I can categorize their classification down to the correct family. I'm constantly amazed at the world that exists below the surface and no two dives are ever the same. This fascination with ocean life is part of the reason I was there. However, if you took all that away, I'd still be in the water for a different reason. I was born a hunter. I've known this from the second I knew what the word meant. Before I began diving I assaulted the shore with rod and reel. Early in life, before I understood the principles behind conservation, I would target everything below the surface and keep whatever I caught. When I began studying the ocean as a young adult, I started to realize that the creatures on the planet were not limitless and that the more we'd take now, the less we'd have for later. At that point, I didn't change who I was, I simply accepted the challenge of targeting specific species and released everything that I didn't want to eat. Understanding conservation and applying it to my outdoor activities essentially honed those hunting skills. No matter how intriguing the underwater world is to me, I will never enter the water without a spear gun. That's just the way it is.

At the surface I spotted a group of large triggerfish darting around a boulder. They appeared to be chasing away anything that came close to their territory and not paying much attention to me. I dropped down behind the boulder and settled on the bottom. I extended my spear gun out in front and waited. Within seconds the largest of the fish raced out to confront the shiny spear tip. When the fish turned broadside I fired. At the surface I pinned the fish to the stringer and moved on down the reef.

I was floating the backside of the reef and was just about to make another dive, when a sharp searing pain made me drop my gun and grab for my right leg. The triggerfish, swinging free on the wire stringer, had drifted up near the top of my thigh and clamped down on my leg with its canine-like teeth. Luckily the fish had grabbed mostly swim trunks, but the pain made me realize it wasn't all he had. I grabbed the fish by the head and pulled it loose. I lifted the edge of my trunks and examined the wound. The sharp teeth did not break the skin, but they left enough of a mark so it looked like I had been attacked by a dog about the size of a shoe.

I dispatched the fish with my dive knife and retrieved my gun from the bottom. At the end of the reef, I swam around to the front side and decided to float with the current and hunt my way back. The rocks at the end of the reef were filled with deep caves and the afternoon sun had cast this portion of the reef into a spooky gloom. The current had just begun to slowly push me down the front side of the rock, when something near the bottom caught my eye. At the base of the reef, about 20 feet down, was a large undulating shape that looked like a discarded piece of carpet swaying in the current. An old rug I could've handled.

The six-foot moray eel was determined to either take my speared fish hanging from my stringer or take a bite out of my leg. He swam straight towards me without fear or trepidation. I quickly started to back-pedal. As I retreated, the dead trigger fish dangled dangerously close to the snapping jaws of the advancing eel. I raised my gun and poked the fish in the face only three feet away. He stopped instantly and just floated there for a few seconds. After that he abruptly turned around and headed back to his lair. Despite the outcome, absolutely no part of me felt brave or victorious. After I floated the front side of the reef, I kicked back to shore drained.

Back at camp I saw Tom messing with his dive gear. He was holding two pieces of rope, one in each hand and looked to be deep in concentration. He told me he was trying to tie his stringer onto his floating tagline using a special Bolin knot. He could've told me he was removing his own kidney and it wouldn't have mattered much to me. I was worn out and a bit defeated from the dive. Despite my disinterest Tom stepped towards me to guide me through the knot-tying process.

Tom and I spent the next fifteen minutes running through the intricacies of the mighty Bolin, but I was not in the mood to be taught anything. After he tied his stringer to the other rope, he grabbed both pieces and gave them a good tug. Satisfied with their integrity he picked up the rest of his gear and headed out towards the reef for a dive.

I took a quick shower and joined the rest of the group. Jo was once again absent and resting in her tent. Her presence on the trip was becoming more of an annoyance than anything else. She hadn't participated in any of the sample collecting, stating that she had no idea how to fish. When the work up station had been set up to process the samples, she had once again found her way to her tent. She apparently had no plans to help out, and no desire to be there.

I grabbed a few Granola bars and sat myself in the shade of one of the palm trees. The dive had been a nice distraction from the science portion of the trip, but we were here to collect samples, not relax. During the dive I had searched just about every inch of the far reef looking for spotted sand bass without luck. Spotties behave a bit differently than other fish. While most of the reef inhabitants cruise around chasing or being chased, spotted sand bass will find a comfortable spot on the bottom and just settle there. I've seen patches of them during dives were they just litter the ocean floor. They all face the same direction, almost like they're waiting for a movie to start. If they were present within the lagoon of Punta Chivato, they would've been easy to spot.

Larry and Danny had taken a few of the students out in the Whalers to fish outside the lagoon. Since they hadn't returned yet, I was hopeful they were catching our target species. I almost got up and started setting up the work up station, but the midday heat kept me planted in the only available shade in camp.

Cheryl came over and sat a chair next to mine. Her beautiful smile and the tilt of her head made my heart skip a beat. We sat there and watched the tiny waves in the lagoon and the frigate birds soar over the boats. When the graduate group began planning the trip and confirming volunteers, I was a bit nervous including Cheryl. We hadn't been dating long and roughing it in Mexico for two weeks was not as romantic as most would think. But her company during the driving stretches and her positive attitude made it all worth it. I can say that right there on that beach, I realized that she was something special and I was glad she was there with me.

Since the prospects of working up fish were slim, those that returned to camp grabbed a chair and squeezed into the shifting shade near the palm trees. There wasn't anything else to do. Tom walked up and sat down in the only chair still sitting in the sun. He didn't look happy. He said he had shot a huge triggerfish and had clipped to his stringer that he had tied to his float line. When he returned to the beach the Bolin knot had come loose and the stringer was gone. He held up his hands to indicate the size of the fish he'd lost. The gap between them was about two and half feet wide. My triggerfish wasn't even half that size. Tom then mentioned that he had found the moray eel at the far end of the reef. He said the thing came right up and let him pet it. I flashed back to my encounter and tried to decide if I had misinterpreted the eel's intentions. It didn't matter. I told Tom I'd seen it too and left it at that.

Just before dark, the fishing group returned with grim news. After spending almost seven hours out on the water, they had caught only six spotted sand bass. Jimmy poured out two handfuls of rubber lures that had several tiny circular bite marks ripped out of them. "We couldn't get the lures down past the damned triggerfish." The leathery fish would race in as soon as the lures hit the water and ripped them to pieces. They're mouths are small and very strong, and they have no problem biting at the rubber and avoiding the hook. Tom and I both reported that the lagoon appeared to be absent of spotted sand bass as well. We collectively thought that the fish we were looking for were probably out deeper. After some discussion, it was decided that if no spotties were collected the following day, we'd pack up and head to our next location and scratch Punta Chivato off the list as a sample station.

At dinner, Eric appeared at the edge of the crowd covered from head to toe to protect whatever live skin he had left. He looked far better than he did a few days earlier. He had taken a pretty heavy basting in Guerrero Negro and most wondered if he'd ever be the same. We even started teasing Larry, telling him that Eric thought the number seven was a color and that he now spoke fluent German. I was glad he was feeling better.

Dinner was eaten around the roaring fire. My triggerfish and a few that the boat crew had caught were turned into some of the best ceviche I had ever tasted. The huge bowl was passed around the fire like a communal meal and was gone by the third go round. Add in a few bean and chili burritos, and as many cold beers on top of the appetizer and most were slumped in their chairs around the fire rubbing full bellies.

After dinner Larry handed out the work assignments for the following day. John and I were ordered back into the lagoon to search the area for spotties. Danny was already getting the boats ready for another day out on the water to search for samples. Mark and Tom would be in charge of dinner duty. They had met a couple earlier in the day that were willing to take them out on their boat to spear fish near the off shore islands. Anyone else was to grab a rod and fish from shore. Everyone had their assignments except Jo. She had once again retired early stating it was too hot to eat and she wasn't feeling good. Larry checked in on her and she said she was just tired. Being tired after not doing anything all day was a symptom. We had no idea what the problem was, but I think most of us started to become a little concerned at that point.

As the beer flowed and the inhibitions came down, Larry and Jimmy became involved in a rather heated argument centered on trash around the campground. The yelling was short lived and the altercation ended in silence and a few surprised stares. As the group sat around in recovering quietness, I thought about how this had been the only real argument among us so far on this trip.

DAY SEVEN

June 27th, 1993

The sweltering temperature had me up with the sun. My shirt was soaked with sweat and I hadn't slept well. To cool myself down I had gotten up in the middle of the night and soaked a towel with water and used it as a blanket. That worked for a little while. Outside the tent, a blast of warm air hit me in the face. It was already in the 90s and the sun was barely in the sky. I could think of no better place to be than in the water looking for fish.

I walked over to the food area and grabbed an orange and a few granola bars and sat by the now smoldering fire. I could see that most of the boat fishermen were up and swarming around the Whalers getting things ready for the day. Eric was bundled up from head to toe and waddling around the boats like a penguin.

Today marked the halfway point of our sampling expedition. As far as fish samples, Guerrero Negro had gone well, while San Quintin had not. Now sitting here on the second day at Punta Chivato, the fishless trend continued. I was pretty sure that the divers weren't going to find some magical school of spotted sand bass in the lagoon. And if the boat crew couldn't get through the triggerfish to the species we needed, we'd have to scrap this location as well.

I guess I had already written this location off. The premise of our research centered on comparing three known southern California populations of the target species, to three Mexico populations. If we couldn't come up with three populations down here, we'd essentially be comparing apples to oranges, and that's never a good idea in science. So far Guerrero Negro was our only solid Mexico population. San Quintin and now Punta Chivato couldn't be considered since we couldn't find the fish we needed. We had one more spot scheduled further south. Even if we collected what we needed there, we'd have to come back to locate a third population. I didn't mind traveling to Mexico, but another trip meant pushing my research timeframe back and spending more time in school.

The fishermen packed everything up and headed out to launch the boats. For a short time, I was the only one awake in camp. Glancing around, I noticed that the place was a mess. There, sitting next to a tent, already in the early morning sun was the blue sample cooler. A twinge of something went through me, but it clearly wasn't enough for me to take action. I almost went over to examine the box, but I opened up another granola bar instead. I knew about all the marine organisms that had been added to the cooler. I hadn't seen anyone add in any additional ice. Whatever ugliness waited inside, I didn't want to see it. I walked over and pulled the cooler into some temporary shade and went to get my dive gear ready.

Tom and Mark had risen early to meet their offshore ride. The owner of the boat told them that he had been out the day before and caught several large fish on the backside of the islands. I was a little jealous, but I really did hope that they'd get a chance to kill some big fish.

Down at the cove the waves were tiny. The day was already hot and I was looking forward to spending it diving. Several splashes caught my attention a short distance from the beach. Something was chasing small baitfish near the inside of the reef and the sun was still too low for me to see what it was.

I eased into the water and slowly floated towards the area. Instantly bait fish swarmed out in front of me darting and moving together. I moved with them and then broke through the school and surprised the predator. A reef coronet fish, close to four feet long, raced into the fray, parting the bait ball. Unsuccessful, the fish glided to the edge of the reef to stage another attack. The bait ball crowded up beneath me and kept undulating to confuse the killer. The coronet fish was not at all concerned with my presence. I watched the long fish prepare for another attack. The swimming stick was scarcely wider than my spear and about twelve feet away. That's when I decided to take him.

I slowly raised the spear gun and followed his movements. I waited for him to turn broadside and fired. The spear hit the skinny fish right behind the head and killed it instantly.

I watched the spear sink to the bottom with the fish attached. I floated at the surface and pulled in my tag line and the speared fish. In less than a minute, I had the long fish strung on the stringer and my gun reloaded.

I continued drifting the reef, occasionally making dives to search some of the smaller rock openings and caves. The area was loaded with life and I decided to gather a few more specimens for the blue cooler and for dinner. I came around the side of a large boulder and surprised a small school of Graybar grunts. One of them was far more surprised than the others and I added him to the stringer as well.

I spent a few hours searching new areas and looking for some food fish. I really wanted to take another nice triggerfish, but I really didn't feel like getting bitten again. Besides, you could tell with all the diving we've been doing in the lagoon, most of the bigger fish were getting wise to our presence.

Around noon the tide shifted. I floated the edge of the lagoon and then found the far side of the reef where I had seen the eel. I poked around a few likely holes and caves but he wasn't home. I wasn't quite sure what I would've done if I had seen him again. I guess it was possible that he was curious enough to greet divers. However, with a face full of sharp teeth and an almost psychotic permanent smile, I'd rather be cautious.

The shadows in the lagoon rose and fell. I had been in the water for most of the day and I hadn't eaten a thing. More importantly I hadn't stayed hydrated and I started to feel it. The dull thump of a headache had been with me for the last thirty minutes and I felt like it was time to get out.

I floated around the outside of the reef for a few more minutes. The wind had kicked up a bit and it was getting difficult to hold my position. I could feel myself being pushed farther out and I was expending more energy than I had. I blew the seawater out of my snorkel and started kicking for the beach.

I walked up on to the beach and sat down on one of my swim fins. I saw another diver out near the outer reef and figured it had to be John. From the position of the sun, I guessed I had spent almost eight hours in the water. The sun took the chill away quickly and the white sand radiated enough heat to warm my entire body within minutes. I lay back and soaked up the warmth with my eyes closed. All I could think about was a cold beer and a burrito.

It wasn't long before I heard John sloshing through the shallows. He tossed his gear next to mine and plopped himself down in the soft sand. While exploring the far side of the inner reef, John had found Tom's stringer. He said it had the biggest triggerfish he'd ever seen strung to it. The other fish had picked it clean overnight so John slid the carcass off the stringer and left it out by the reef.

Camp was bustling with activity when we returned. Paul was getting dinner going and the fishermen were once again trading stories. Only a few had been wondering where John and I had been most of the day. Mark and Tom had returned from the pleasure boat trip and both had speared nice Dorado.

Tom's fish was stretched out on one of the coolers as he put away his gear. The dorado was close to twenty pounds and had a small spear hole right behind the gill plate. I sat there waiting for the story, but Tom didn't seem to be in the mood to talk. I could tell something was bothering him. Finally, after a bit of coaxing, he briefly described his day to me.

Apparently Mark had been extremely demanding during the boat trip. The couple had graciously agreed to have Mark and Tom on board as guests, and Mark had acted like an ass. If areas weren't holding fish, he told the captain he wanted to move. This went on for most of the morning. Towards the end of the day, when it became apparent that Mark was not going to share any of his fish with the boat owners, Tom jumped back in and speared another Dorado for the hosts. Back at the dock, Mark had thanked the couple, got off the boat and left. Tom hung around afterwards and helped clean up the vessel and then gave the owners some money for fuel. Tom didn't know Mark very well and he said the entire day was a huge embarrassment. He told me later that he'd never dive with that guy again.

I gave Tom his lost fish stringer and that seemed to cheer him up a bit. Although I never did see the triggerfish he had shot, the description from John was enough for me to consider it worthy of mentioning. I told him it was a nice fish and he appreciated that.

The fish I had speared were odd species and weren't considered dinner fare. Larry mentioned that if we encountered anything unusually, he wanted to see it. Besides, I couldn't imagine you'd get much of a meal out of a coronet fish. My fish were headed for the blue cooler.

At the cooler, I secretly wondered what waited inside. I was pretty sure that no one had taken the time to add more ice, let alone clean the thing out. I slipped the fish off the stringer and took a deep breath and held it. I quickly opened up the cooler lid, tossed in the fish and quickly shut the top. The faint wisp of decay caught my nose. I had gotten enough of a look inside to notice that the last remaining ice in the thing had melted away and nothing new had been added. A slight sting of guilt quickly pulsed through me and then just as suddenly was gone.

The fishermen had been pushed off the water by the afternoon winds. They had only landed two spotted sand bass before that. This was not good news, but it wasn't totally unexpected. We had sampled Punta Chivato hard for two solid days and had only scrapped up a handful of target specimens.

We hadn't even come close to the number we needed. I knew that if we didn't collect more specimens, I essentially didn't have a graduate project. That evening it was decided that we were done with Punta Chivato.

After dinner the group settled in around the fire. Once again Jo was absent. She had spent the entire day resting in one of the tents and I think it became clear to everyone that something was definitely wrong with her.

Milton Love, a visiting professor from Santa Barbara and a good friend of Larry's had come on the trip with his thirteen year old son Elan. When he heard that Jo was still acting sickly, he decided to go to her tent and find out what the problem was. Greg and I followed. The three of us entered the large tent and found Jo sprawled out on the top of her sleeping bag. We instantly realized that something was wrong. Jo had spent the entire day inside the tent, but she looked like she had sunburn. Her face and arms were dark pink and her hair was matted to her face. She looked as if she had a high fever. She also had small sores covering her face and arms.

Almost instantly I knew what it was. Jo had chickenpox. She sat up and looked at her arms. She mentioned that a few weeks before the trip, she had done some babysitting for a neighbor whose two small kids were just getting over the virus. I looked at Greg, and Milton looked at me. We all shook our heads as if answering the unspoken question. I had been unfortunate enough to contract the dreaded pox as a freshman in high school. I can't say I remember the episode fondly.

We left the tent and quickly took inventory of the group. As it happened, Mark was the only one who hadn't tangled with the chickenpox earlier in life. He quickly developed an obsessive routine of washing everything several times and staying as far away from Jo as possible.

Larry met with Jo and she assured him that she wasn't feeling that badly, and she wanted to continue. Her fever had dropped considerably by evening and she actually started eating again. Drawing from my own images of being fourteen and completely covered with little sores, it appeared to me that Jo had contracted a rather light case of the virus and possibly the worst was over.

That evening we relaxed and talked about the day. Jo was feeling better and even decided to join us for a beer around the fire. This of course sent Mark to bed early. Mike, Cheryl and I grabbed our lawn chairs and a several beers, and headed down towards the shore to watch the large dark clouds of an approaching storm system. Deep inside the huge shapes, lightning popped and glowed as the clouds slowly moved their way over the gulf. Nights like that make you hate your television.

We lined the shore and let the warm gulf waters lap at our bare feet. There was no better way to end the day. Mike reached over and handed me another beer. He picked one out for himself and opened it. "How long of a drive is it to Magdalena Bay?" Mike asked.

I took a long swig of the ice cold beer and let it cool my guts. "I think it's about seven hours," I said. It would definitely be the longest leg of the trip and I wasn't looking forward to the drive.

With the nonstop drive down from the States and the lengthy run from Guerrero Negro to Punta Chivato, Mike had spent most of the trip driving. And since Mark and Mike had driven down together, he was essentially stuck with his travel partner and his quirky habits. He was a desperate man.

"So, any chance of you and me switching travel partners?" Mike asked.

I took another long swig of beer for dramatic effect and looked out over the gulf. The choice to either travel the peninsula with a pretty lady, or a neurotic guy that didn't like to listen to the radio was an easy one. For fun, I let Mike think I was really undecided. I took another long swig of beer before I answered. I declined the offer.

Mike was clearly frustrated with the answer, but he knew what it would be before he asked the question.

We finished our beers and let the sounds of the gulf make us sleepy. We all turned in early and drifted off to the sounds of distant thunder and small waves lapping at the shore. As I slept, turmoil and microscopic rage were plotting. Festering like a filthy demon in the darkness, the blue cooler was about to rear it's stinky, ugly head.

DAY EIGHT

June 28th, 1993

The sound of the wind whipping the canvas of my tent shook me from a dream that I couldn't remember. The storm had advanced during the night. I unzipped the tent and stepped into the humid air. It smelled like rain. Paul, the cook, was busy packing up all the cooking essentials for the next leg of the trip. A bowl of fruit and a box of granola bars were the fare for the morning meal. I grabbed an orange and a few bars and walked over to the dead fire to eat.

The cloud system had traveled our way during the night and it was now looming large and hovering over our camp. Dark clouds blocked out the sun and cast the beach into a dismal gloom. Even with the cloud cover it was humid and hot. A dull flash blinked deep inside the storm clouds and I began to count. I reached six before the low frequency rumble reached my ears. I figured that within the hour it would be raining at the beach. Fortunately the next stop was further south and we'd be driving right out of the approaching storm system.

The weather flashes and rumbling thunder had many in camp awake and packing up early that morning. Mark, anxious with the approaching weather, had quickly rolled up his tent, clothes and sleeping bag into a dirty ball and stuffed it into the back of Mike's car. He now sat in the passenger seat nervously waiting for the rest of the camp to get moving.

I had already started packing Cheryl's car with all our gear for the next leg south. With dive equipment, a tent, sleeping bags and clothes, the little compact car was just about loaded to capacity. I was pulling the spear gun in through the inside when I saw Larry across camp standing by the blue cooler. He was standing, hands on hips, looking down at it. I froze. The top of the cooler was not completely shut, the result of either excess internal gas production or negligence. I watched as Larry bent down to open the lid. As soon as he tilted it back, he turned his head quickly. He took a few steps back and even from where I was I could tell he was angry.

I decided that I had seen enough. I hunkered down in the back seat of the car, pretending I was fixing a broken seat belt. As I focused on the fake repair, I secretly wondered whose name he would call first. I knew for a fact that I was on the short list for blame, but I was feeling lucky. "TOM!" Larry bellowed, standing over the cooler.

Larry spent the next ten minutes tracking down anyone that might be responsible for the decaying infractions in the blue cooler. Since many had added items to the soupy contents of the box, Larry could really be angry with our entire group. However, since the larger grunt and the clams and crabs made up a majority of the biomass, John and Tom took the brunt of Larry's anger.

Later on Larry found me and strongly suggested I head over and assist in the clean up. I walked over to where the two were busy scrubbing scum and decay from inside the box. I stopped and looked down at them in disgust, shaking my head. "Man," I said, "is Larry mad at you guys!"

The plastic bag filled with the genetic samples from Guerrero Negro had popped open and tiny vials floated in the gray water. This was obviously a contamination problem. Since we'd pass right by the seaside town on our way back, it was decided that Carrie and Tom would make a stop there to collect more fish blood to replace what had been soaking in decay.

We finished packing and cleaned up around camp. Mark had spent the last half hour sitting in the passenger seat of Mike's Toyota, eyeing the approaching storm. We made sure that everything was tied down and slowly drove out of camp. Just before we started out, I glanced back and noticed an old dog meandering over towards the liquid pool that we had poured from the rancid cooler. I wondered if the next group of campers would find the camp hound expired.

A short time later we were back on Highway One and heading south towards Magdalena Bay on the Pacific side of the peninsula. The visit to the gulf had been refreshing but unproductive. We were unable to locate enough spotted sand bass in Punta Chivato to consider it a viable sample location, and just like San Quintin, we ended up dropping it from the research study.

The main highway south of the Bay of Conception rides the coastal hills of the peninsula and parallels the gulf. For almost four hours the brilliant blue of the Sea of Cortez was visible off to our left.

Around noon we pulled off the main highway into the town of Loreto for supplies. Since the food van was all packed up, we found a small restaurant and decided to sample some of the local cuisine. Before we even sat down, Jo once again stated that she wasn't feeling well. Sick himself of the constant complaining, Larry took her aside and discussed the options available to her at this point in the journey. He explained that we could take her to the local airport and put her on a plane back to the States at her expense. Or, she could stop her complaining. Being a student and relatively poor, she decided to stop complaining and joined us for lunch. As far as the illness, she appeared to have gone through the worst of it in Punta Chivato.

After stocking up on food items and finishing lunch, we got back on the highway and headed south. We bordered the gulf for a few more miles and then angled slightly inland for the cross to the Pacific side. The storm system we had left in Punta Chivato was barely visible behind us, and it looked like the hot gulf air was keeping it stalled in the Bay of Conception. Ahead were clear skies.

Just before dark we reached the turnoff to Magdalena Bay. We wound our way through the town and located a small public launch ramp near an old fish processing plant right near the shore. We pulled through the open gate and drove right to the edge of the cool Pacific of Magdalena Bay. A few yards down the shoreline were two primitive shade areas constructed of plywood, old lumber and palm tree parts. Way off to our left was a building that marked the edge of the parking lot. The cement structure ended at a high pier that sat fifteen feet above the water and stretched another one hundred yards past the edge of the bay. Way off to our right was a primitive launch ramp carved into the hard dirt of the shore. The shallow incline was lined with thousands of clam and oyster shells used for traction when launching a boat.

In the dwindling light we grabbed our fishing rods and lined the shore. As lures splashed in the shallows and the reels clicked, the inhabitants of Magdalena Bay soon made their presence known.

Both John and Greg hooked up on their first casts and landed a couple of very fat spotted sand bass. Larry's nephew, Doug, caught the first short-fin corvina of the trip and Cheryl added a broom tail grouper to the species list. All on shore were either hooked up or dealing with a flailing fish bouncing at their feet. It was like the fish were stacked up near the shore waiting for us.

Mike, Larry and I quickly grabbed our rods and joined the rest of the group down at the shore. Larry's lure hit the surface on his first cast and was grabbed as it sank. He fought and landed the two-pound spotted sand bass. I was slow-swimming my little rubber lure back to shore, when it was grabbed five feet from the rocks. I set the hook sharply and played the unseen fish for a few minutes before finally landing an almost legal California halibut. And Mike, well he just kept on casting.

As the group made casts and unhooked fish, it wasn't difficult to notice that one member of our party wasn't doing much beyond swatting at the air with his rod. Mike had made a dozen casts without getting a bite and this had not gone unnoticed. I had taken to tossing the fish I was releasing right in front of Mike. They would land with a splash and usually startle him. Both Larry and Danny made less than encouraging remarks on Mike's fishing ability as he struggled. And Mike just kept on casting. I was just about to suggest that he hang it up and head back to camp to get us some drinks when he finally set the hook on a fish. *"That's right!"* he yelled, as he fought his first fish of Magdalena Bay.

After a seesaw battle that lasted a minute or two, the defeated fish resorted to its last line of defense; it puffed up like a balloon and floated at the surface like a two-pound sponge. The laughter started before the fish was landed. Mike had caught the beach ball of the sea. The puffer fish frantically flapped its tiny pectoral fins and grinded its buck teeth against the hook sticking in the skin near the mouth. The fish wheezed as air and water escaped it as it rolled in the dirt near Mike's feet. "Sounds like it has asthma," I said, laughing at both the fish and Mike.

We spent the last hour of daylight fishing and enjoying each other's company. At one point I looked down the shore and six of us were hooked up. Since we had arrived late to Magdalena Bay and were not prepared to work up any of the specimens, all the fish were returned to the bay, we hoped to be re-caught the next morning. We cast into the setting sun, and I began to realize that this was the trip of a lifetime.

DAY NINE

June 29ᵗʰ, 1993

The next morning the sound of the van engine revving and shells cracking under the tires had me peeking from my tent. Paul was at the wheel and trying to back the boat trailer down the shallow makeshift ramp and into the water. The van was in perfect alignment. The boat, however, was angled almost parallel to shore. Paul pulled away from the beach to try again. One chance is all you get with Danny. He parked his cigarette on his tackle box and walked up to the driver's side of the van. He opened up the door, and without a word Paul got out of the van and Danny got in. With minimal effort and in less than thirty seconds, Danny had the boat floating in the shallow bay. He parked the van and walked back to the ramp to get his gear. He leaned down, picked up his cigarette, grabbed his tackle box and sloshed into the water, shoes and all. He boarded the Whaler and dropped the engine into the water. He started the outboard and looked back towards Paul, squinting through his own cigarette smoke. Danny backed up the boat, and just before he turned to leave, he smiled the smile of the competent back at Paul.

Larry and Danny were once again taking the Whalers out to fish off shore. The evening before Larry had instructed us to catch and work up what we could from the beach.

After breakfast, I headed down to the shore with a rod and a bucket. The water inside the bay was like glass, the tiniest of waves lapping a rocky shoreline. I turned the bucket upside down and sat on it. As I rummaged through the tackle looking for a lure, a small splash caught my attention 30-feet from where I sat. A silvery flash blinked once under the surface and then was gone. Immediately a second splash in the same general area had me fumbling with my gear.

I made a cast towards the splash. The lure sank, but never made it to the bottom. The line twitched once and then stopped. With the bite, I set the hook and felt angry shaking at the other end. After the fight, the first spotted sand bass of Magdalena Bay was curled up in the bottom of the five-gallon bucket.

I looked down at the fish. Its gill plates were flared out and the fish seemed to dare me to grab it. The body was far rounder than the other spotted sand bass we had caught in Mexico and it looked to be feeding well. As I added more fish to the bucket, I noticed that this shape was fairly typical of the spotted sand bass of Magdalena Bay.

For a while I had the fishing action all to myself. I had caught thirty of the target species in about an hour and half. The fishing action hadn't diminished from the previous evening and the fish didn't seem to care what color lure you tossed to them. I had already worked through a dozen different colors of rubber and the chewed remains of each floated in the fish bucket with the fish.

Rods in hand, John and Cheryl joined me at the shore. For the next few hours we enjoyed almost non-stop action on the spotted sand bass. We also caught halibut, grouper, trigger fish, corvina and puffer fish from the same beach. It was one of the most amazing fishing days I had ever experienced.

With three buckets completely full, we stopped fishing and headed back to our camp to begin the work up. A valuable lesson learned from our experience in Guerrero Negro, we had decided to pace ourselves and stretch out the science portion of the work into two days.

We set up the work tables under the shade of the primitive structures near the shore. In a few short minutes, all the implements of science were set up and ready for the first spotted sand bass of Magdalena Bay. During the preparation, I watched John grab the data book and park himself in front of the work up table. I reached over and grabbed the data book and handed it to Cheryl. In Guerrero Negro John's napping habits had caused a serious data gap and I had no intentions of letting that happen again. John agreed to help fillet the fish and to assist in keeping the science machine moving.

We all took our spots and I started moving each fish through the process. John had pulled out his tape player and turned up the volume on one of his heavy metal selections. Almost immediately we fell into the efficiency of field data collection and repetitive fish processing for science. We avoided any data mishaps this time and John was filleting the last of the fish by mid afternoon. During the clean up, Cheryl informed us that we had worked through 137 fish during this last session and that all the data was in order.

An hour later, Larry and the rest of the boat crew returned from their day out on the water. They were festive and excited as they motored into the shallow bay. They beached the boats and motioned for us to come give them a hand with their gear. John and I walked down to the shore to help them offload two large coolers, heavy from the day's efforts. We grabbed the plastic handles to drag them off the boat but the weight kept them anchored to the bow. John and I looked at each other.

With more effort we yanked the cooler from the bow and carried it over to the work area. I remember thinking that if that cooler was filled with spotties, there must be over two hundred inside. I opened the larger cooler and felt somewhat relieved. Lying in the bottom were fifteen silvery fish about two feet in length. A few had died with their mouths open, and each displayed two very impressive canine-like teeth sticking out of the roofs of their mouths.

John quickly opened up the other cooler and found the same species stacked inside. The fish were short-fin corvina, fairly common for this area. The uncommon feature of the fish was their size. After a few minutes of leafing through the pages of their reference books, Larry and Milton were confident that with the specimens they had collected that morning, they could document a species size increase for the fish.

The fish in the coolers were silvery sleek and mean looking. And as soon as I saw them, I wanted to catch one. Danny and Paul grabbed the nearest cooler and started pulling fish out to fillet. Larry had intended on letting us loose the next day to fish off the boats. However, since we hadn't reached our target number, that wasn't going to happen. Fishing for fun would have to wait.

With a few hours of daylight left, a few of us grabbed rods and again headed down to the beach. If we caught most or all of the remaining samples we needed, we'd be that far ahead during the following day's work up.

Shortly before sunset, we had all four sample buckets overflowing with the target species and a rough count of almost 150 fish. We iced down the buckets and kept them covered with plastic to remove any temptation from the camp dogs. We even went as far as to set up our assembly line that evening. As I tested the batteries on the digital scale, I thought about how much I really wanted to catch a short-fin corvina.

DAY TEN

June 30th, 1993

The next morning I awoke to a silent camp. Both Whalers were gone and I didn't feel good about it. We still had over one hundred fish to work up here in camp and the guys with the boats would have no idea when we'd be finished. To top it off, this was our last full sampling day. It would take us two full days to drive nonstop back to the states and we were supposed to head back early tomorrow morning.

After breakfast I walked over to the fish buckets. They all appeared to have survived the night. I kicked one of the buckets and a few fish at the top came to life and struggled briefly before their confinement settled them down. It was time to get started.

By now the work had become very familiar to us all. With the familiarity of the tasks came efficiency. By noon we had managed to finish up ninety fish. Even though the species and jobs were the same, you could feel that we were getting faster.

We took a short lunch break and snacked on cheese and crackers. Afterwards we cleaned up the area a bit and jumped right back into it. We still had approximately fifty fish to work through and I was ready to have the fish processing done. I turned up the radio a bit, put on a fresh pair of gloves and picked up another fish.

In the world of high quantity fish processing, the last twenty fish are always the toughest. You know that you're close to being done, but twenty fish can take twenty minutes or an hour. Really, the only time you feel like things are moving well is when you are completely finished.

As we were reaching the bottom of the last sample bucket, I stretched my sore back and noticed some movement inside Larry's Bronco parked nearby. I could see two little mischievous heads in the back seat doing something. Both Eric and Elan had been absent from camp for the last twenty minutes, and while I was relieved that I now knew where they were, for some reason I could just tell that they were up to no good. The cloud of white smoke rising from the back seat pretty much confirmed this.

Mike, being closest to the car, yanked the door open and found two surprised little boys playing with a lighter and an old cookie box. By the time that slight distraction was cleaned up, I placed the last fish on the cutting board.

The last spotted sand bass was stretched out, measured and weighed. I announced the length and weight to Cheryl, the data recorder. I glanced over at her and she gave me a smile that made me blush. The fish was an average specimen for the location, and after I cut it open, I declared it a female. I pulled the swollen ovaries out of the fish and using a pair of scissors, snipped the tissue towards the front. As she had done over five hundred times before, Carrie was at my side holding a sandwich bag open containing a small numbered piece of waterproof paper. I dropped the tissue in and returned to the fish for the final sample. I cut the auditory canal, just behind the fishes head, and cracked it open on the slimy edge of the measuring board. I held the flip top fish up to Greg and he carefully pulled the aging bones from the fishes head. He dried them off and dropped them into a coin envelope with the same sample number written on it.

I held the last fish up and looked at it. The spotted sand bass is a beautiful fish. Almost maroon in color dorsally, the ventral area turns to a crème color. Possessing a pair of golden eyes and several faint stripes made up completely of small brown spots. Females have a yellowish chin that becomes more prominent during the breeding season and males have a whitish chin. They are aggressive in their daily habits and pure energy at the end of a line. And while I was absolutely sick to death of the species at that point in time, I was elated to have collected what we had, and equally excited to be working on spotties for my graduate research project.

Later that day Larry let John and I take one of the Whalers out to fish for fun. We each caught several of the toothy corvina and both enjoyed the freedom of running the boat, free from sampling and authority. During the run out to the mangroves, I remember tilting my head back in the rushing wind and feeling the sun heat my face. I could've driven that boat like that for the rest of my life. While most of the days of this trip run together for me now, the one that stands alone is that day out on the boat fishing with John.

As with most things we look forward to and enjoy, they come and go way too quickly. The radio crackled once and Larry's voice came on telling us it was time to return to camp. The trip back to the states was going to be long and nonstop. Larry wanted both boats out of the water and on the trailer for an early start the next day.

I often think about this part of the trip. I suppose I didn't truly realize the expedition was just about over. If I had, I would've taken the time to thoroughly enjoy the short boat ride back to camp.

When I pulled the Whaler into the small bay, I could see that the other boat was already out of the water and resting on the trailer. Everyone within sight was busy packing up gear that we wouldn't need for that evening. The sight of camp being rolled up for the last time depressed me. And I could see from the pace of the pack up, I wasn't alone. The duties were forced and even from the boat, I could see the mood was somber. During the trip when we had packed up before there was always excitement in the air. We had no idea what to expect during the next leg and that blind adventure is what we craved. It was that urge of the unknown that developed during the trip that made me realize I had stepped through to a new career and more importantly a new way of life. And now I could see that the celebration was over. The trip was done. It was time to head home.

I mark this trip as one of the most memorable and important in my young life. It was most assuredly a trip of a life time and you usually remember those. I forged some lifelong relationships and feel almost certain that I will remember and interact with most of these individuals for the rest of my life; both as friends and colleagues. I also firmly believe that this trip marks where I stepped through the door of science.

We had successfully sampled two of the four stations chosen, and both Guerrero Negro and Magdalena Bay would become crucial locations in the research of several graduate projects for the university. Both San Quintin and Punta Chivato were dropped from scientific consideration due to the lack of samples and replaced with Los Pulpos, the university's research post located on the gulf side of the peninsula.

As life moves on, I think often of this trip. As young adults, we were in that gray area of life, somewhere between education and career. While we all had aspirations for the future, we enjoyed living in that moment as field biologists. Traveling to a new land and not knowing what lay ahead was all part of the adventure and we had all adapted. For us, it truly was a trip of a lifetime.

After we graduated, we all went our separate ways. I still stay in close contact with almost everyone involved on this trip and consider all of them my good friends. I was even lucky enough to find my best friend and wife in Cheryl. And even though at times, Larry had to tug hard on the reins to guide our unruly group, I will always consider him a good friend and mentor first and my academic advisor a distant second. Words can't express my appreciation.

And while it took a year or two for me to convince Danny we were friends, I can think of no one else I would've rather had on this adventure. My early participation as a field assistant and researcher were shared with Danny and he always made sure that I knew that hands down, the most important person on the boat was the guy that could fix it. I have never forgotten that. You are welcome on my deck anytime.

Mike and I don't see each other as often as I'd like, but I do give him all the credit for convincing me to leave the engineering field to give field biology a try. His presence on this trip was invaluable in so many ways. My travels through the training of science would not have been the same without Mike.

I feel extremely fortunate that the path I've chosen has brought me into contact with these people. It isn't always the destination that makes the memories. For me those that travel at my side are what make the trip. And about this trip, I can say emphatically that I would not have wanted to be anywhere else at the time.

The Research Vessel (RV) *Yellowfin*. This vessel was my home away from home for almost four years during the summer research trips and university contracts. (SCMI)

The sturdiest boat I've ever piloted. The 18 foot Boston Whaler that never let me down. If I knew where she was today, I'd buy her. (Tim E. Hovey)

This photo marks the first day of my new career path. Despite my eagerness and enthusiasm, I had absolutely no idea where this training would lead me. (Carrie Wolfe)

Free diver and volunteer, Steve Redding collecting data on the back deck of the *Yellowfin* in 1991. This ended up being the last photograph of Steve I would ever take. (Tim E. Hovey)

The typical living conditions of a field biologist. It's easy to see why this life style doesn't appeal to everyone. (Tim E. Hovey)

Enjoying some down time on the deck of the RV *Yellowfin*. (L to R) Holly Harpham, Carrie Wolfe, John Smith and the author. (Michael Franklin)

A group of students working up fish samples at the wet lab. No matter what sampling occurred out on the open ocean, all the important dissecting and data collection happened here. (Tim E. Hovey)

Me just happy to be back on board the *Yellowfin* after the shark encounter. As a diver, it's best not to think what may be swimming out there just beyond the eerie gloom. If you did, you'd never get in the water. (Larry G. Allen)

Tom and the white sea bass he speared off the coast of Catalina Island the day we ran into the shark. That smile says it all. (Tim E. Hovey)

Chief Engineer, Danny Warren on the back deck of the *Yellowfin*. Danny convinced me that hands down the most important person on a vessel was the guy that could fix whatever broke. His sense of humor and honest opinions made every voyage out to sea more fun than work. (Tim E. Hovey)

Relaxing and enjoying good company on the back deck of the *Yellowfin*. At the end of the day, part of my job was to make people laugh. (L to R) Greg Tranah, the author, Dr. Larry Allen and Danny Warren.

A view of a male spotted sand bass I'm all too familiar with. By the end of my graduate career, I had cut up over 2,000 specimens of this species. (Tim E. Hovey)

The field crew hunkered down next to the wind break at San Quentin, Baja California, Mexico on the first day of our two week trip down the peninsula in 1993. (L to R) John Smith, Greg Tranah, Ron Klaver, the author, Andy Barbarina and Cheryl Baca. (Carrie Wolfe)

The three wise men and the clown relaxing at Guerrero Negro, Baja Mexico. I owe most of my career success to these three individuals in some way or another and I feel beyond fortunate that they let me tag along. (L to R) Dr. Michael Franklin, the author (background), Dr. Larry Allen and Chief Engineer Danny Warren. (Cheryl Baca)

John and I collecting samples off one of the huge cement fishing platforms at the shore of Guerrero Negro. We took great pride in being able to target specific species and catch them for research projects. (Mara Morgan)

Getting ready to assess the near shore for science at Punta Chivato, Baja California, Mexico. (L to R) John Smith, Greg Tranah, Cheryl Baca and the author. (Carrie Wolfe)

The quiet shores of Magdalena Bay. We fished from the rocky beach to the left, and worked up the samples in the primitive shack to the right. This was our last stop down the peninsula.(Tim E. Hovey)

Me (left) and John Smith holding two nice short fin corvina on the shore of Magdalena Bay, Baja California, Mexico. This was the last sampling day during the two week trip down the peninsula. This adventure would have been far less enjoyable without John. (Cheryl Baca)

Celebrating our arrival at Oke Landing, Baja California, Mexico. (L to R) Greg Tranah, the author, Holly Harpham, Tom Grothues, Dr. Larry Allen, Cheryl Baca, Andy Barbarina and John Smith. (Danny Warren)

Setting up the fish processing work station at Oke Landing. Our sample team would process hundreds of specimens in a single day, no matter where we were. (Tim E. Hovey)

Part of the sample team after a successful day of field collecting at Oke Landing searching for gold spotted bass. (L to R) Tom Grothues, John Smith, the author and Andy Barbarina. (Cheryl Baca)

Fishing the deep waters and glassy conditions off the coast of Oke Landing for gold spotted bass. This was definitely the fun part of science. (Tim E. Hovey)

The morning of the Northridge quake. Thanks to a failed marriage, I had moved out of this apartment complex two months before it collapsed. The white address numbers are attached to a second story balcony. The first floor collapsed, dropping the second and third floor twelve feet, killing sixteen people. (Tim E. Hovey)

The brand new university parking structure destroyed the morning of the Northridge quake. I will never forget the sound of the revving engine deep inside the tangle of cement and steel. (Tim E. Hovey)

My beautiful wife Cheryl in La Paz, Mexico with a 40 pound *Yellowfin* tuna she caught. After countless adventures and primitive conditions, she remains at my side. I am beyond lucky. (Tim E. Hovey)

Dr. Larry Allen, friend, mentor and advisor in that order. Larry holds a firm spot as the second most influential person in my adult life, right behind my dad. (Tim E. Hovey)

A RECORD MEAL

Once we returned from the Baja trip, John, Greg and I stayed very busy working on the samples we collected in Mexico. We had returned to the States with a dozen, 5-gallon buckets filled with fish and fish parts. All the samples were in support of our projects and they needed to be sorted, catalogued and prepped for the lab portion of our research. And this was only half of our samples. Our research proposals included collecting specimens from several southern California populations of spotted sand bass as well, and those samples had yet to be collected.

Despite our focal research projects and the huge mound of work we still had ahead of us, John, Greg and I had other work duties as part of our graduate status. We were still involved in several research contracts during the summer months, as well as any side projects Larry may have been involved with. If it involved collecting samples, net hauling or fishing road trips, we were usually pulled from our projects to assist. And to be honest, these short trips and relatively regular contract requirements were a welcome break from the lab tedium.

One of these side projects involved a life history study of the gold spotted bass. This species of fish was a member of the same family as our target species, but occupied far deeper water than the coastal spotted sand bass. Unlike other members of this fish family, not much current information or research had been done on it. And finding patterns or the lack of patterns within a species family, was a huge step in defining its life history and drawing a solid conclusion on the habits and behaviors of the species. Larry was understandably interested in adding another closely related species to the work load. Before the summer was over, we were once again headed down the Baja peninsula to collect as many gold spotted bass as we could catch.

We had received reliable information that local fishermen in the town of San Felipe, located in the northern part of the Gulf of California, were regularly catching gold spotted bass near the offshore islands. A few of Larry's fishing buddies had confirmed the presence of the species at fish markets in the popular tourist town. After referencing maps and checking on roads, Larry decided to point his sampling crew to a desolate stretch of rocky shoreline called Oke Landing. The islands were a short five miles off the shore of the abandon fishing village and would be an easy run for the small work boats.

After a full day of driving, the road group pulled into Oke Landing and began setting up camp. If you weren't looking left towards the gulf as you drove south, you'd easily travel right passed the abandoned fishing village. The shoreline was completely covered with small, flat rocks, forming a slight cove. Two large rocky cliffs, liberally covered in bird droppings, bordered each side of the beach and cast the camp in shadow when we arrived. The rocky beach angled upward sharply and leveled off a short distance from the road. Several very old cement foundations sat near the flat area, the only remnants of human presence on the desolate beach. Within minutes of our arriving, several dome tents were being set up on the relatively flat foundations.

The rocky islands were visible a short distance offshore. To a bunch of field biologist in training, they were more than inviting. A few of the travelers had launched small skiffs at the small coastal town of Puertocitos, 30 miles to the north of our rocky camp. Leaving their trucks and trailers at the town, they ran the work boats south and met up with us at the abandoned cove.

The all too frequent patterns of these research trips involved the sample collection and the sample work-up. The students got to participate in the hook and line sampling, which we all really enjoyed. Unfortunately, our student status also meant we were responsible for processing the samples as well. And we all secretly knew that the more fish we caught during the collection, the more work we'd have at the end of the day.

After we set up camp, Larry gathered us all around to explain what we needed from this sample station and what we'd be doing. Summer days here were blazing hot and we needed to be out on the water early the next morning. The two boats would be loaded with anglers and we'd fish around the islands until we located the gold spotted bass. Once we located the fish, we'd catch as many as we could before it got too hot. The plan was then to have the students work up the catch back on the shore in the cooler part of the late afternoon.

Early the next morning, we loaded up the student boat and headed for the offshore islands. The species we were after lived 250 feet below the surface and regularly congregated near ocean ridges during the summer. After about an hour of dropping heavy lures laced with squid to the rocky bottom, we finally started collecting the gold spotted bass. I radioed Larry's boat that we had found the school and soon both Whalers were drifting a hundred yards apart filling buckets with samples. We spent the next several hours catching our target species and just having a great time. This was without a doubt the fun part of science.

The gulf water was like glass and it was beyond hot. There was not a hint of wind, and while it was nice to fish these flat calm conditions, the extreme heat had me wishing for a breeze.

A short distance away, usually always within earshot, Larry's nicer boat floated motionless in the flat conditions. Both small skiffs were Boston Whalers, but that is about where the similarities ended. Whenever we headed out to sample, the students were always lumped together in the older, beat-up skiff and the friendly group of captain, engineer and professor got to fish in the nicer boat. It was the unspoken rule of the trips and none of us really cared. And while we always enjoyed the freedom of running our own skiff, the fatherly tether to our boat was never more than a shout or a radio call away.

We floated on the flat calm sea dropping our offerings to the reef below. Once the lure hit the bottom, we'd engage the reel and give the silvery lure a few twitches to entice a bite. We never had to wait long. Specimens of the fish in all sizes would grab the lure almost instantly and more than once we'd hook two fish on the same lure. They were then hauled to the surface, pulled from the hook and tossed into the sample bucket.

For a few hours the student boat was lost in the pleasure of just fishing. We enjoyed the freedom of running our own boat and the success of catching what we were after. For a short time we were just a bunch of friends spending a great day out on the water.

I glanced over to the other boat and noticed Larry motioning us over. I had seen this before and I knew what it meant. The lowly graduate students were done fishing. It was time to get to work. "O.K. boys, it looks like we're done fishing for the day," I said.

John, Greg and Andy reeled up their lines and stowed the fishing gear in the bow. During the sampling we had filled almost three five-gallon buckets with fish. I moved the buckets to the stern in anticipation of picking up a few more from Larry's boat. I started up the motor and headed the short distance to the other skiff.

I pulled up next to Larry's Whaler and killed the engine. Since the collecting had gone so well, Larry suggested we head back and get a start on the work up. Danny pulled two full buckets of fish to the side and loaded them into our Whaler. With almost five full buckets and a rough estimate of over 100 fish to work up, we wouldn't be finished until dinner time.

Back at camp we offloaded the fish buckets and placed them under the shade covers we had set up earlier in the day. The shade did nothing to cool us from the 100-degree heat and the windless day was going to make processing the fish miserable.

John and Greg started setting up a row of gear on the tables in the work area. First in line was a measuring board complete with a slime drain hole. After we had worked up hundreds of fish at a time, the small hole allowed the build of slime to ooze out the back and was definitely a necessity. Next to that was a digital scale. Next came two cutting stations; one for removing the gonads so a reproductive strategy for the species could be established, and one for removing the ear bones for aging purposes. After the cutting stations came sample storage. A single chair was placed in front of the work area, reserved for the data recorder.

The data collecting requirements for the gold spotted bass were exactly the same as the data collected for the spotted sand bass. Establishing and comparing life history information among different closely related species begins with collecting absolutely everything you can when you have the fish in hand. One of the most important aspects of the life history of any fish is the size and age in which it reaches maturity. Maturity is defined as the first season where the species successfully leave offspring. Combining size and age data with the point of maturity will establish a size and age at first maturity, probably the most important aspects of fisheries management. So the measuring and weighing of the fish, the tissue collecting and otolith analysis all tie into defining a complete life history for the species.

After twenty minutes of preparation time we were ready to get started. Each fish was carefully measured, weighed and then cut open in an assembly-line type operation. Once science was finished with each fish, the larger ones were filleted and saved for dinner. The smaller fish were fed to the ever-growing crowds of shore birds that had instantly appeared once we had started the dirty part of science.

As is often the case with species that have received very little scientific attention, no one was really sure just how big these fish got. We had finished up two buckets of fish and were just starting the third when I started to notice that the specimens from this bucket were significantly larger than the others. The bucket also contained several empty beer cans indicating that the collectors were having a lot more fun than we were now and were probably still out collecting. This was one of the buckets we had picked up from Larry's boat.

Each fish removed from Larry's bucket seemed to be larger than the previous one. With every read out of the weight, Cheryl, the data collector, would declare the current specimen to be the largest so far. The grams on the scale continued to rise, 1,000, and then to 1,500. A few smaller fish moved through and we figured we had finally worked up all of the larger ones. Then we found the bottom of the bucket.

There, sitting in several gallons of putrid water, was the tail of a fish far larger than any we had worked up that day. I carefully removed the specimen and placed it on the measuring board. It eclipsed the length of any of the previous fish by more than six inches and it had a girth that almost made it look like a different species. It looked twice as large as any of the other fish. The giant was transferred to the scale and we all quickly gathered around in anticipation. There really isn't much excitement in the repetitive processing of fish for science so you really have to take your thrills where you can get them.

The digital scale flickered once and then started to climb. The display finally stopped at 2,996 grams or over six and a half pounds of fish. A true giant when you consider that most of the specimens we had already processed hovered around two pounds. We admired the fish for a few brief moments, worked it up and then passed it on to the fillet table. In the world of fish work-up, size really means little. The fleshless carcass was tossed into the pile with the rest of the remains and quickly forgotten. Had we known at the time that we had just handled the king of the gold spotted bass, we may have had a brief moment of silence.

Larry's boat cruised into the tiny cove as we finished up the last bucket of fish. He jumped from the still moving craft as it hit the shore and headed our way. He casually searched the buckets we had already worked through clearly looking for something. I guess he had been the angler that had landed the monster gold spotted and wanted a picture of the fish. Unfortunately, there is only one destination for a sample once we get a hold of it. What was left of Larry's giant fish was now being picked clean by the western gulls over on the carcass pile.

With a handful of smaller fish left to process, I felt confident that Larry's fish was by far the largest we had worked up. Cheryl leafed through the data sheets and let Larry know the weight of his giant. That seemed to satisfy him. Despite his lofty scholastic status and rich research career, part of Larry was just a fishermen looking to land the biggest fish.

The next few days we continued to catch and work up gold spotted bass until we had all we needed. Once the magic sample number was reach, we stowed all the sample gear in their work boxes and sealed the preserved specimens in five-gallon buckets with lids. With the fish work finished, we spent some time exploring the offshore islands for fun. The mood was high as professor and students alike flourished in the glow of another successful sampling trip.

A few weeks later we were fulfilling another one of our graduate student obligations by working the fisheries booth at one of the popular tackle and boat shows held every summer in the Los Angeles area. We spent time talking to the public about research and what our fisheries program was attempting to do. This was usually followed up with suggestions of research donations or the sale of a fisheries shirt that displayed all the sport fish we routinely worked on.

A large aquarium trailer sat nearby filled to the top with seawater and containing several live specimens of each of the sport fish we conducted research on. The aquarium belonged to the university and always drew a crowd. The public would frequently gather around the huge fish bowl and most seemed very interested in learning about the local fisheries. Since the show primarily targeted boaters and anglers, the large aquarium was an easy icebreaker when it came to talking about sport fishing. And to be honest, without it, we'd just be another booth asking for a hand out.

During one of the breaks, John and I decided to head out and explore some of the other displays at the show. Most were selling expensive fishing gear and brand new boats, and while there was little chance we could afford any of it, we did enjoy looking.

Once we walked through the tackle booths, we found ourselves standing on the far side of the main area in front of a vendor that produced high-quality fish mounts. Dozens of examples of his work decorated the wall and stretched for almost 100 feet. John and I walked the display and tested each other on the identification of some of the more obscure species gracing the wall. We made our way to the center of the display and there it was.

Hanging from the display wall, directly behind the vendor and occupying a place of pride was a beautiful replica of a large gold spotted bass. I was surprised to see the species represented on the wall since I hadn't heard much about the fish until we started working on it only a month before. Directly beneath the mount, in a polished frame, was an International Game Fish Association world record certificate. It verified that this specimen was an all-tackle size class world record for the species. The biggest fish of its kind ever recorded. Printed on the certificate, in bold type was the weight of the fish. The fish weighed six and a quarter pounds, a full quarter pound smaller than the fish Larry had caught in Mexico.

John and I just looked at each other and started laughing. I thought about how we had chopped up that big fish into fillets and fed it to the field research group in Baja. The day ending with our sitting around the fire eating fish burritos as the sun set over the peninsula. With little more than satisfied burps, we had consumed the largest gold spotted bass ever recorded.

Eventually we found our way back to our booth. Larry and one of his fishing buddies were just leaving to check out the rest of the displays around the convention. Like a couple of nervous school girls, John and I just couldn't keep our new secret in. Just before they left, we strongly suggested they head over to the north end of the auditorium to check out the fish mounts.

My shift at the booth was over and I decided not to hang around. I gathered up my stuff and headed for the exit. I laughed at the thought of a mount of the second largest gold spotted bass in the world. The carcass of the largest ever recorded sitting in a pile of fish skeletons somewhere on a beach in Baja. I couldn't believe that we had caught and eaten a world record.

As I reached the convention exit I heard a loud, somewhat manly scream from the direction of the display booth. I assumed Larry had found the world record fish. I felt a little bit of pride, knowing that I had handled, processed and hacked up the largest gold spotted bass in the world. I guess it would be like eating the last panda on earth and not knowing it.

LESSER LESSON

One of the benefits of traveling into Mexico and collecting our own samples for research is that we frequently caught or observed other species south of the border. Often these non target fish would either be released or added to the fillet bucket for dinner. However, when new students came into the fisheries program at the university and expressed an interest in working on species we may have encountered during our collecting trips, the veterans of the Mexico sampling knew where to get them.

Mara Morgan was a new graduate student and interested in looking at the genetic differences between California halibut in southern California and those off the Baja coast. Students new to the program usually came in with an idea of what species they wanted to work on. However, Larry's general interest for his lab was fisheries life history comparisons. This could be in depth studies involving all aspects of the species, like the research that had been conducted on the spotted sand bass and gold spotted bass. Another acceptable project would be a comparison of genetic differences or similarities between populations of the same species that may be geographically separated.

During our sampling trips into Mexico, we would occasionally catch halibut at three of our sample stations along the coast. We didn't catch a lot of them, but we had seen them at Guerrero Negro, Los Pulpos and Magdalena Bay. The year before, during a dive across the bay in Guerrero Negro, I had seen several juvenile halibut scooting around the shallows. We had also caught a few from the shore in all three sampling stations. If Mara was interested in collecting specimens from Mexico, we had a good idea where to start.

Our fisheries group didn't really mind collecting specimens for other fisheries projects. Usually the student or project leader would conduct most of the work up themselves, leaving us to simply collect specimens. Occasionally we'd help out with the dissecting and tissue collection, but for the most part our duties were to collect the samples needed. Collecting specimens for genetic projects was even easier. All that was required from the fish was a species confirmation and a small sample of blood. To make sure the samples were taken correctly and the blood wasn't cross contaminated among individuals, the student conducting the research almost always did this portion of the processing themselves.

Within weeks of finalizing her project, the sampling group was scheduled for a road trip down the Baja peninsula to collect fish. Mara's project required fifteen samples from each Mexico location to compare to the California population. With four samplers targeting just halibut, I was hopeful we could get the job done.

I pulled the black Suburban up within feet of the old salt processing plant of Guerrero Negro. The headlights lit up a faded piece of graffiti on the side of the building indicating that we had been there before; "CSUN was here," was scrawled in red chalk near an old doorway. I had left the message the year before during our spotted sand bass sampling. Seeing the college logo gave me a sense of familiarity and security for some reason. This was to be our first stop during our week long sampling trip. If we were able to collect what we needed here at Guerrero Negro, we'd continue down the road to Magdalena Bay.

Greg, John and Mara started unloading their tents and sleeping bags. I went back to check on the boat. The skiff had taken a pounding on the way down and at one point we had to pull over and readjust the tie downs. They had come loose while traveling an exceptionally bad piece of Mexico highway. I checked the tow hitch. The bow strap attaching the boat to the trailer was tight and secure. The transom support rod was firmly in place, keeping the engine from bouncing around and damaging the transom. The hull was beyond bulletproof, solid and seaworthy as usual. The four full gas cans were securely tied down, as was all the sampling gear we had carried with us in the skiff. Fishing rods, tackle, dive bags and spear guns were secured in a neat bundle at the bow. Despite the degraded road, the skiff and all its contents looked to be in fine shape.

We set up camp near our old campsite and cooked up a quick dinner of burritos and beer. John's CD player was blasting heavy metal music off the walls of the old processing plant and in the light of the single lantern the scene looked like a bad horror movie. We sat around on coolers and overturned buckets and enjoyed the evening.

The following morning, issues with the boat kept it on the trailer. The engine would catch, cough clouds of thick white smoke and then die. For several hours John and I fiddled with parts and checked what we could. Switching out gas tanks finally solved the problem.

With everything finally in order, we eased out of the tiny harbor and headed for a stretch of sandy beach on the far side of the lagoon. Once I pulled into the channel, we checked to see which way the tide was running. Our plan was to head to the front of the drift and have the current slowly push us along the beach just off shore. With all four of us fishing and not having to worry about maneuvering the skiff, it would save us time and increase our chances of catching what we needed.

The tide was running to high, so a massive amount of water was rushing into the huge bay. For us to take advantage of the water movement, we needed to run to the mouth of the bay and start fishing there. The tide would push the skiff along and once the drift was finished, we could run the boat back to the mouth and do it again.

About fifty yards from the beach, and at the head of the current, I cut the engine. The water was flat calm but moving. The beach looked familiar. The year before several of us had taken one of the skiffs to this area and enjoyed a calm beach dive. During that dive I saw several juvenile halibut scooting around the shallows. With the tide conditions, it was the best place to start sampling.

We spent the first morning drifting the shallows of the sandy beach with light line and green, curly tailed grubs. Within minutes we started collecting the target species. On the first drift we caught three juvenile halibut and by noon the total was up to seven. All the fish were about 12 inches long and looked to be about the same age class.

Each fish was tossed into the large cooler, half filled with sea water, as soon as it was brought aboard. With three of us still fishing, Mara would collect the blood samples she needed from each specimen. A small section of gill tissue was excised from each fish while it was still alive. The sample was placed into a small vial filled with a chemical buffer. The vial was marked with all the specific data, placed in a small lock box and stored under the center console. After the samples were collected, the rest of the fish were tossed into the cooler to be filleted later.

In the early afternoon the wind picked up, making drift fishing with light line and light tackle impossible. The stiff breeze pushed the Whaler quickly down the shore and I knew we were done fishing for the day. We headed back to the bay, happy with the day's efforts and glad we were able to catch what we had. After we stored the blood samples in our sample cooler with dry ice, we filleted the small halibut that had given their lives for science. In the shadow of the old salt plant, we fried up the fish fillets for an early dinner and chased them down with cold beers. For desert we opened up John's ample supply of Pop Tarts.

As the sun dipped towards the horizon, Mara, Greg and John retired to the coolness of the gulf for a swim. I grabbed my rod and headed for one of the large fishing platforms at the end of the beach. I walked out onto one of the huge cement cylinders that lined the shore of the bay. Four of these structures stood side by side on the beach, just in front of the old salt processing building. Large chunks of thick and heavily scared rubber were bolted to the sea side of the towers with huge rusty brackets. The tangle of metal brackets that connected the cylinders together and extended into the water, created the perfect hiding place for fish.

I leaned against a 3-foot section of I-beam, rusted and worn by the elements, and started fishing. I made casts and hopped the lure along the bottom on auto-pilot. I thought about our first full sample day in Guerrero Negro, and about the hassle with the boat. Despite our late start, we had caught half of what we needed from this station. Good, but not good enough. We had one more day here to collect what we needed, and I was confident that without any other delays, we could accomplish the task.

I watched a brown pelican practically float only a foot above the calm water, a dozen yards from where I stood. It suddenly drew its wings back, dropped its head and pierced the bay water with a calm splash. Several bait fish scattered as it lifted its head to situate and swallow the unlucky few it had captured. "Well, at least you're catching something," I told the bird. After almost 30 minutes of fishing and at a time of year with the temperature right, I had yet to even get a bite.

I grabbed my tackle and decided to head back to camp. A young Mexican man, who had been fishing on the other side of the same platform with 40 pound mono wrapped around a beer bottle, walked over and started a conversation just as I was leaving. We each struggled through the language barrier. After a few minutes, I was finally able to decipher that a massive salt spill the week before had wiped out thousands of fish near the shore. Even though the processing plant no longer produced and packaged salt, a few of the locals still made a living collecting and selling the mineral locally. The young man then pointed to an eddy forming under one of the platforms. There in the golden light of the setting sun were dozens of dead and bloated fish swaying in the tide.

He turned and walked to the edge of the platform and got his primitive gear ready to fish. I watched as he bent down and started smashing small clams on the worn cement to use for bait. I looked down at my tackle bag, cluttered with countless lures and lead heads. I walked over and gave my new friend a handful of tackle and wished him luck. Just before I rounded the building to our camp, I heard some excited yelling from the platform. I turned to see the Mexican fisherman holding up the thick line with a fish dangling from the end. He was smiling and waving my way. I smiled and waved back at him. While I was too far away to see what he was using for bait, I like to think he was using one of the lures I had just given him to catch the fish.

I walked into our make-shift camp in the shadows of the old warehouse. The group was gathered around the fire warming up. Greg and the gang had enjoyed their swim, and they were right in the middle of describing one of the creatures they had seen when I returned. John had spotted the large fish in the shallows and couldn't quite identify the flat, shark-like creature. He described it as being about 3 to 4 feet long and looking like an angel shark. The group discussed the sighting and after I heard several descriptions, I wasn't convinced that John had spotted an angel shark this close to shore and this far south. However, without seeing the creature myself, I couldn't offer any suggestions on what it was.

The four of us sat around the fire that night, talked of life and solved half the world's problems. With another full day of sampling ahead of us, and still another collecting station further south, we let the fire die down and then retired to our tents.

We slept like rocks and rose with the sun. Minus the engine trouble, we were out on the water hours earlier than the previous day. We motored to the same spot we had drifted the day before and started fishing. By noon we had only managed to land four more halibut, putting the total for this station at eleven. All the fish we were catching were about the same size.

In researching population dynamics using genetics, the size of the specimen is really insignificant. Unlike life history studies where size and age classes are crucial in establishing important aspects like growth rates and size and age at first maturity, with genetics, all we needed was blood. When the research question involves the comparison of genetic patterns between different groups of the same species, as long as we can collect enough blood for a sample, we really didn't care how big they were.

It didn't take long for the rising sun to burn off the marine layer. I could feel the day getting hotter and I decided to shed my sweatshirt. I made a cast and let my lure sink to the bottom. I placed the rod butt between my knees to hold it and began to remove my sweatshirt. As soon as the article of clothing completely surrounded my head, a strong tug almost pulled my rod overboard. I quickly regained my composer and grabbed the fishing rod. Twenty feet below us in the crystal clear water, attached to my lure, was the beast.

"That's it," John said. "That's the fish I saw diving yesterday!" The group stopped fishing and stared into the clear water to get a good look. The large, shark-like animal had a broad head, a slender tail and was struggling against the current. The moving water and its flat head made it almost impossible to reel in. I tried to raise the rod and bring it to the surface, but it wasn't moving. While I could plainly see it wasn't an angel shark, I still needed a better look to figure out what it was. I did the best I could on light tackle to lift it's huge pancake-like head to the surface, but I could feel I was losing the battle. The ray was just too big. About ten feet from the boat, the large fish shook its head strongly and the small hook pulled out. The sand-colored ray floated back to the bottom and out of sight.

After the sighting, I tended to the small halibut we had caught that morning. I clipped a gill filament from each fish and placed a blood sample into a plastic vial. I tossed the whole halibut into the cooler and felt the first hint of the afternoon breeze wash over the boat. We had just completed a drift of the sandy area without any luck. I knew with the impending wind, we needed to increase our odds if we were going to collect the remainder of our samples. It was time to get into the water and find them.

John and I untangled our spear guns and float lines, and got the gear ready. The current was still running strong and I figured we could use that to our advantage. As we suited up, I told Greg to drop us off at the beginning of the drift at the far end of the bay. Once we finished drifting through the area, he could pick us up, and conditions permitting, we could make another dive. While John and I were diving, Mara and Greg would continue to fish from the boat and hopefully between the two groups, we'd be able to catch the rest of the samples.

Greg ran the Whaler up the beach as John and I got suited up. He eased the throttle back near the mouth of the bay and close to the beach. He turned the Whaler around and pointed the bow in the direction we had just come and turned the engine off. Since we were separating, we discussed the plan thoroughly. I didn't want us to drift too far apart in case one needed the other. Both John and I agreed to stay out of the main channel and relatively close to the beach. Greg said he'd try to drift with us and stay close. With the plan firmly in place, John and I eased over the side into the warm water at the head of the drift.

The visibility was excellent and the sandy bottom was easily visible 25 feet below. I kicked in towards shore a bit and found an area that looked promising. I floated motionless at the surface and noticed the bottom moving by very slowly. The current speed was perfect for hunting.

I drifted over the sandy floor, looking for any inconsistencies that could be halibut. After investigating a few spots up close, I started to wonder if I'd be able to see the camouflaged fish from the surface. I glanced over and saw John cutting in closer to shore and shallower water. Further down the beach I saw Greg and the Whaler fishing the end of the drift. During the drift, I noticed several small, regularly placed depressions on the sandy bottom. From the surface I could see that something appeared to occupy each little depression. I took a few quick breaths and then headed towards the bottom to investigate.

I leveled off at the bottom near one of the small sandy potholes. The dents in the mud were about fifteen inches in diameter and only a few inches deep. I edged up to the nearest depression and there, inside the hole pretending not to be seen was a single seahorse. I watched as the small animal furiously beat its pectoral fins to maintain its position in its tiny foxhole. As I drifted the bottom, I carefully visited several additional holes, each holding an adult seahorse defending its little territory. I gently picked one of the tiny animals up. It instantly curled its little tail around my finger in an attempt to drag me back to its lair. I shook the small fish from my hand and it drifted back to its little home. As I kicked to the surface, I realized that not many people get to see the amazing things I've seen under water.

At the surface, I continued drifting down the beach and searching for halibut. Other than the heard of sea horses and a few spotted sand bass, fish life in the area was pretty sparse. Half way to the boat I spotted John drifting the shallows. I noticed that he didn't have any fish dangling from his stringer either. We still needed a few more halibut from this station, and even though we weren't seeing any, the drift dive was relaxing and enjoyable.

At the end of the drift, John and I swam over to the boat and climbed back in. The wind was picking up a bit, but I figured we had time for one more dive. Greg ran us back to the front of the drift and we started again. I took a slightly different depth back, but I still wasn't seeing much.

I had just reached the surface after a dive, when I spotted something lying in the sand on the bottom. My mind was convinced it wasn't a halibut, but the hunter in me had to get a closer look. The flat, skate-like animal was partially buried in the sand and motionless. From the surface, I could see darker rings, like a bull's-eye centered on its back. It was about three feet long and definitely resembled a smaller version of what I had hooked and lost earlier that day.

As soon as I saw the skate partially buried in the sand, I knew I was going to spear it. Since we didn't know what it was, I figured we could take it back to camp and use our field books to identify it. We were also quite used to eating just about anything we captured when field sampling in Mexico. Unless the field guides stated it was inedible or poisonous, our group would easily carve dinner out of it.

I adjusted my mask and made sure my spear gun was in order. I took a few deep breaths, dipped my head beneath the surface and headed for the bottom. I came up slightly behind the fish, and leveled off about three feet from the bottom. I slowly raised the gun and aimed at the concentric markings centered on its back. I straightened my arm and pulled the trigger. The spear found its mark.

The shaft went through the middle of the fish and stuck it to the muddy substrate. I dropped down to the bottom and knelt next to the struggling ray. I reached down and grabbed the spear and yanked it from the sea floor. A cloud of mud and silt erupted beneath the fish and just briefly, I lost sight of the ray. I then reached around and grabbed the spear point on the other side of the speared fish, capturing the beast between my hands. The ray kicked a few times, but the fight was all but over. As I turned to kick to the surface, the fish slid down the spear and made contact with my left hand and most of my left arm. I vaguely remember seeing the eyes of the ray sink into its body. It then energized its kidney shaped, battery-like organs and doused the surrounding area with a strong pulse of electricity. With the steel shaft and the idiot that held it, the circuit was complete.

My first sensation at the bottom of the ocean was the feeling that nails were being driven into my knuckles. My grasp on the spear became involuntarily tighter and the muscle contractions in my biceps jerked the speared electric fish close to my face. As I looked up to the surface, my view became clouded in a tunnel-like vision. I felt my teeth grind as the current clenched my jaw muscles, cutting through the plastic mouthpiece of my snorkel. The last thing I remember, there on the sea bottom, was feeling an insane surge of panic rush through my body.

I don't remember surfacing or dropping the spear. I only remember knowing that I had been shocked to near death underwater and it was my own fault. I felt angry, embarrassed and very disappointed in myself. Still shaking, I spotted Greg in the Whaler a short distance away and motioned him over. I floated there for a few seconds still groggy from the jolt. While I recovered, I looked down at the scene on the bottom. The dead electric ray, still skewered by the spear, lay in a tangle with the tagline. The floating line snaked to the surface and bent slightly to the current. Next to that, shaken loose from the jolt, was one of my swim fins lying in the mud. Surrounding the gear, the area displayed spastic and frantic scuff marks in the sediment, clearly illustrating my electric struggle for life. I closed my eyes, and weakly floated there waiting for the Whaler.

In less than a minute Greg was at my side with the skiff. "Are you O.K.?" he asked. I said nothing as I grabbed the side of the Whaler and pulled myself inside. I reached down and grabbed my floating spear line and pulled my gear aboard. When the spear and ray arrived at the surface, I carefully and quickly unscrewed the spear point and pushed the dead fish off the shaft and watched it sink. I dropped all my gear in the bottom of the boat, leaned back on the bow step and let the sun warm my battered body.

John appeared at the side of the boat before the ray hit the bottom. As he watched it sink, I saw him take a dive to grab my fin. When he surfaced, he grabbed the side of the boat. He hefted himself up so only his arms and head were visible from where I lay. "Hey, why didn't you save that fish so we can figure out what it is?" he asked, tossing my fin into the boat. I flexed my numb hands and my eyes remained closed. "No need," I said, from the bottom of the Whaler. I don't know if the electrical jolt helped my memory, but I suddenly remembered everything I wanted to know about the lesser electric ray. Although why they labeled it 'lesser' was beyond me. I suddenly became very thankful that I hadn't run into the larger specimen I had hooked earlier.

As I recounted my story to the rest of the group, I realized that my left hand was still a bit numb and that both of my biceps were sore. Greg piloted the skiff to a small cove near our camp, and out of the afternoon winds. The four of us fished hard the rest of the day and scrapped out the remaining samples. The next sample station further south had gone well and after spending a few days relaxing in Mexico, we headed back to the States.

Once I got home, I searched through a few of my fish books to educate myself on the lesser electric ray. I was a little surprised to see that these types of rays are actually rated according to the voltage they can produce. The "lesser" electric ray was rated to deliver a DC voltage of between 48 and 54 volts using its muscle cells to store the charge. Each cell is oriented plus to minus to amplify the charge each cell produces. The bigger the ray, the more electric cells they have, thus the larger the charge. I read the description again. I'm no electrician, but I was pretty sure the jolt I had received was equivalent to what was needed to light up a small city.

I closed the book and thought about the trip. Even though we had some issues with the boat and I had almost been killed by a fish, it had been a successful collecting trip. I thought about the good friends that had assisted on the adventure. I knew that I would know all of them for a long time. I thought about that stupid ray and the far stupider human that had decided to tangle with it. I remembered the scene as the dead fish sank to the bottom. I was angry at myself for messing with the fish, and far angrier for wasting it. Although I don't think I could've convinced anyone in our group to fillet the thing for dinner. I thought about passing out underwater and how I don't remember a huge chunk of the whole encountered. It was just another reminder to me that if you make a mistake under the water, it can be your last.

A RUDE AWAKENING

In January of 1994 our fisheries group was in high spirits. Not only had we survived our first semester as graduate students, we had all made excellent progress on our research. All the sample trips had gone well and only six months into our graduate careers, we had collected all the specimens we would need for our individual projects. Even Larry was impressed with our progress. However, he always made sure that we knew we weren't quite done yet.

The three of us had worked hard to neatly categorize, label and store our samples safely for later analysis. Greg's genetic samples sat neatly organized in several small sample boxes in a storage refrigerator in the genetics lab. All the vials containing bits of tissue required for his entire study could easily fit into a small suitcase. John's samples were also small. The tiny ear bones he had removed from the juvenile fish's head for aging purposes were hardly larger than a grain of rice. Those were stored in coin envelopes and all the data for each fish was scribbled on the outside. John's entire sample load for his project could be carried around in a large lunch sack.

My project required larger pieces of the fish. To figure out the spotted sand bass's reproductive strategy, I needed to collect and preserve both lobes of the gonad tissue. Depending on the size of the fish and the time of year, these chunks of tissue can get quite large. Fortunately, since the maximum size of this species is around four pounds, a set of gonads could easily fit into a sealable sandwich bag.

However, with size comes bulk. While the research samples of my two colleagues combined could fit into carry-on luggage, my sample bulk would require the use of a large van to move around.

I had collected over 2,000 individual tissue samples for my project and stored them in twelve, five-gallon buckets. Larry and I had decided to store the buckets in a vacant lab on the third floor of one of the older science buildings. He was concerned that fumes from the tissue preservative could cause problems. I would eventually excise small tissue chunks from each gonad lobe and run them through the extensive chemical protocol of tissue histology. After the sample was treated, I could slice very thin sections of the tissue and mount them on microscope slides. Since I had to take a sample of each lobe, and each fish had two, my 2,000 specimens would eventually become 4,000 specimen slides. I would then examine each slide, one by one, looking for tissue structures that would identify the reproductive strategy of the spotted sand bass.

Our group was well on its way to mapping the complete life history of the fish we were studying. We were also equally as far at fulfilling university requirements for our graduate degrees. In short, we were way ahead of the game. As we gathered at the university pub to celebrate our research progress and the end of our first graduate semester one mid-January afternoon, Greg, John and I didn't have a care in the world.

At 4:31 AM the next morning I was violently thrown from my bed in my rented room above a garage in the San Fernando Valley. Lying on the floor in the pitch black and hearing my stuff fall all around me, I remember thinking that the townhouse next door had exploded. I can say without hesitation that I have never been woken up so violently before or since.

The 6.7 magnitude Northridge earthquake lasted for 20 seconds and killed 57 people. Sixteen of those in the Northridge Meadows apartment complex, a place I called home only two months earlier. It was later determined that the epicenter was approximately 10 miles beneath the earth's surface and about three blocks from where I lived. The Northridge earthquake was the worst natural disaster to hit the valley in 80 years.

A few hours after the quake, as soon as it was light enough to drive, I drove through the chaos of blazing gas mains, toppled buildings and shocked residents. For some reason I pointed myself towards the university and the science buildings. I guess I really had nowhere else to go. All the phone lines were down and calling in or out of the valley was impossible.

I pulled into the large parking lot east of the school. On the right sat the brand new five-story parking structure, still smoking from its collapse. The huge cement columns were grossly bent and the entire pile was folded over on itself like a fallen cake. Deep inside the crumpled mass was the sound of a revving engine. The high-pitched growl was eerie and unsettling. The thought of someone trapped or dead in a vehicle deep in the center of the huge blocks of fallen cement made me realize for the first time that people likely died during the quake. We would find out later that it was an unmanned generator used to light the structure until permanent lights could be installed. Still, I will never forget the helplessness and dread I felt hearing that engine deep inside the tangle of cement and steel.

I parked away from any high structures and made my way to the science lab. A large clock on the side of the university store sat frozen at 4:31 AM, the exact moment the facility lost power due to the violent, early morning quake. As I walked towards the fish lab, I noticed smoke coming from one of the older science buildings. I walked over to investigate. As I rounded the corner between buildings two and three, I noticed fire and smoke pouring out of a single window on the third floor of the science lab. Instant dread and panic shot through me as I carefully counted over the placement of the windows and re-confirmed the floor where the blaze was. Thick black smoke poured from only one window of the building, the window to the lab where I had stored my valuable research samples.

For the next several weeks we weren't allowed to enter many of the buildings on campus, especially the science buildings. The amount of possible biohazards created by the quake was enough to keep anyone without a respirator and authorization from entering. If the Biohazard signs, armed guards and caution tape weren't enough, campus security passed around a list of potential hazards that could be encountered after the quake. Besides the obvious threat of gas leaks, building dirt and caustic dust from smashed computer monitors, more dangerous potential lay behind the doors of the university chemistry labs.

The countless chemicals shaken from their cabinets were now mixed into possibly deadly combinations on the laboratory floors. Drop down a level and the threats, albeit tiny, were now alive. The air-borne pathogens that were stored in cold storage for medical research purposes in the virology lab were likely no longer cold or stored. That bit of information had me holding my breath whenever I needed to be around that building.

Several of the science buildings required cold storage in one form or another for storing sample specimens or even the occasional whole animals for dissecting purposes. Freezers of every shape and size littered each floor of all three science buildings and all were usually stuffed to capacity. Now, every single one of them had been without power for weeks after the event. A quake half the magnitude of the recent temblor would have had little trouble tipping all the appliances over, spilling their now thawed cargo into a stench-filled ooze on the floor. And in and among the vile contaminating sludge, that I was convinced could only be cleaned up by workers in hazmat gear using huge snow shovels, were my valuable samples.

Greg and I had used a moving cart to transport the twelve buckets of samples to the abandoned third floor lab in early January. I wheeled the cart to the back of the lab. To keep the samples off the floor and out of the way, we placed the buckets on the uncluttered workbench in the back, next to two stand-up freezers. Each bucket and each sample inside was labeled with my name, Larry's name and our laboratory number. Now, a few weeks after the disaster, I was convinced that those buckets had spilled their valuable contents all over the lab floor and they were now mixing with the primal ooze creeping out of the tipped freezers.

In the following weeks, with the cleanup of all the labs in full swing, I had pretty much conceded that my research samples were gone and I'd have to start all over. I had even looked into planning another trip to Baja to re-collect what had been lost. But whenever I'd start gathering the paperwork together to move forward, I would think about the immense amount of work I had already done, and what I would now have to redo. It would almost instantly take the wind right out of my sails. I can say that even though I was excited about being in graduate school and working on fisheries research, the months after the quake had me feeling pretty low.

In early March the fish lab was re-opened for use. All the damaged computers and monitors had been replaced, or were in the process of getting replaced. Larry had used FEMA money to make sure that our lab continued to have the tools for us to move forward on our projects. For weeks we would hear a knock on the door and the university delivery staff would ask us to sign a clipboard and then wheel in the newest item to the lab. It had become an almost daily occurrence.

I was sitting next to the door when the knock came. I opened the door to another delivery person. He was holding a clipboard and simply asked if this was Dr. Larry Allen's lab. I nodded and he asked if I'd follow him. I figured we had another piece of gear being delivered, so I politely followed the gentlemen. We dropped down to the basement floor and walked down a short hallway. We made a left turn and stopped. A large moving cart had been placed to the side of the hallway and it was covered with a dirty tarp. The delivery guy grabbed the edge of the tarp and peeled it back. There, neatly stacked, were twelve white buckets filled with my fish samples. "Are these yours?" he asked. I almost started crying.

He told me that the hazmat crew had entered the science lab on the third floor and began cleaning up the mess. Almost instantly several of the crew came upon my samples spread out over the floor. Seeing the labels and the labeled buckets, they carefully refilled the buckets and set them off to the side while they cleaned up. Once the cleanup was complete, they forgot about the buckets and simply left them in the lab. When the university had come in to evaluate the cleanup and approve the room for re-use, they noticed the buckets and asked the delivery staff to find the proper owner. And here they sat.

I feel very thankful that I survived the Northridge quake. I think about those who lost their lives, some in a collapsed three-story apartment building I had been living in just two months earlier. And while my education at the university continued, in my opinion, the university itself never fully recovered while I was there. Many of my graduate classes had to be held in temporary buildings and several labs were closed down permanently.

I have many to thank when I think about my graduate days. Larry was an unwavering symbol of support and guidance through my entire time at the university. I can't imagine that I would have achieved the level of success in my career or in life without him. My close friends and graduate colleagues were also a valuable part of my experience in the program, and I stay in close contact with them to this day. However, I believe the biggest thanks go to the members of the hazmat team that realized the value of what had been spilled before them in the old science lab on the third floor. I frequently think how pivotal the decision was to pick up the samples and not just scoop them up and toss them in with the freezer ooze. It is for this reason that I dedicated an entire paragraph to the cleanup team in the acknowledgement section of my thesis. And I never even knew their names.

A SHARK DIVE

As life gradually returned to normal after the quake, I stayed busy in the fish lab. Even though I wanted desperately to cart all my research samples to my apartment where I could keep an eye on them, Larry convinced me that the fumes from the preservative would probably kill me. "And then where would we be?" he added, throwing up his hands. We finally decided to store the samples in the wet lab next to the office.

A few weeks into summer of that year, my good friend Carrie Wolfe dropped by the lab for a visit. Carrie was another graduate student of Larry's, but she conducted her research at another facility and we rarely got a chance to see her. We talked a little about her research, but mostly just caught up. She mentioned that a gentleman who owned a shark diving charter boat had stopped by the dive shop she worked at part time. He was interested in hiring Carrie for his next charter to add a scientific slant to the two-day shark trip he offered his customers. She had stopped by the lab to see if anyone wanted to come along and assist. Without hesitation I jumped at the opportunity.

While all of us in the fish lab were studying fish, and not sharks, my boyhood interest in fisheries and the ocean had started with the toothy creatures. I would spend all my free time as a kid reading about sharks and could identify just about any species that was worthy of being depicted in magazines or books. As I got older, my interests expanded to ichthyology and the biology of marine creatures, but my childhood interests of sharks never waned. If Carrie needed someone to assist on the dive charter with shark information, I was happy to help.

A few weeks later Carrie and I were waiting on the boat dock just before sunrise with our gear. The dive catamaran was going to stop by the harbor and pick us up prior to picking up the first charter. While we waited, Carry filled me in on what we'd be doing. On this first trip we would be required to give a brief slide show on the dive boat to the divers, focusing on sharks and their habits. That I knew. What I apparently was confused about were the sleeping accommodations and which boat we'd be staying on. I guess I just assumed that we'd be staying on the shiny, new dive boat with everyone else. I was wrong.

Carrie told me that there was already another boat anchored out in the channel that would serve as our home for the next two days. This boat would also be used to establish a chum slick to attract sharks to two large diving cages they had tied off to the stern. The chum boat, one far less prestigious and a lot older than the state of the art catamaran, was already anchored off Santa Barbara Island. If sharks were sighted during our stay, we'd contact the main dive boat anchored at the nearby island. They would then race out to the chum boat and let those interested climb into the floating cages to enjoy the experience. It appeared that we were not only onboard to give a shark seminar, but also to assist in manning the chum slick.

While the owner of the dive company understood that the main species we'd likely encounter in this area were blue sharks and makos, he knew that the real draw for his customers would be to see Mr. Big himself, the great white shark. Since the divers on the charter were also going to be diving some of the beautiful areas near the island, it was decided by the dive boat owner that our boat was not to contact them unless we specifically attracted a great white shark to the dive cages.

While the chum boat rolled and swayed in the channel, the paying customers would get to spend their time in relative comfort and dive the calmer, protected areas near the island. Our crew would spend their time grinding up fish guts and dumping the bloody slurry over the side. Other than the seminar we were slated to give at the end of the trip, our jobs were pretty simple; we were to chum the channel for sharks and only contact the dive boat if we positively identified a white shark in the slick. I had to think that even a Kansas farmer could positively identify a white shark.

I stowed my gear, and during the five-hour trip we were able to enjoy the comfort and amenities of the brand new dive vessel. It had a full galley, plenty of room for up to twenty tank divers and a deck Jacuzzi to keep the divers warm between dives. The catamaran was sleek and handled the slight swell to the island flawlessly. The cook even prepared and served us a nice breakfast as we made the crossing. We ate scrambled eggs and drank orange juice as the dual bows cut through the sea. I was so lost in the sparkle of the new vessel, that I completely forgot we weren't staying on it.

About mid-day we pulled up to the back of the chumming boat. It looked more like a floating shack. The vessel was roughly 40 feet in length and looked to be built sometime in the 1940's. It was in rough shape and without a doubt she had seen far better days. A thick anchor line stretched from the bow winch, down into the water and disappeared into the depths. The bow rails were rusty, missing parts and looked more treacherous than safe. The upper deck was housed in faded Plexiglas held in place with rusty brackets. The side window had enough bird crap on it to completely obscure any viewing, inside or out. The wood trim was bleached white and splitting, and instead of adding a touch of sleekness and class to the craft, it illustrated areas to avoid unless you liked splinters. The tarnished brass screws in the aged teak were either missing or partially sticking out, sharp and waiting for flesh. Two captain's chairs, one missing the back and both oozing stuffing from the cushions, were bolted to the top deck far too close to the edge. On one side of the tiny back deck, trash and various mechanical parts were haphazardly tossed into the corner. On the other side, taking up the rest of the deck room, sat a large shark cage that could easily hold three people. Inside the cage, oozing gray slime and looking stinky, sat four large white coolers. The second cage was in the water tied up to the back and floating several yards off the stern. I glanced over at Carrie and she gave me a nervous smile. I think we were both surprised the boat, which was to be our home for the next few days, was floating.

The engineer on the catamaran tossed four large, white bumpers tied to cleats over the side to keep the two boats from touching. He then walked back to the stern and pulled the swim step door open and stood in the opening waiting for the vessels to get close. A deckhand on the chum boat kicked a rusty stern door open so that we could step across from the catamaran to the work boat. The captain on the work boat saw us approach and never left his chair. He just scowled in our general direction and then turned to work on some made-up job near the boat's control panel. Working the twin engines, the captain of the catamaran floated us close enough so that Carrie and I could step across without the boats touching. Which I'm sure was a huge relief to everyone aboard the pretty boat. The instant transition from luxury to poverty was hard to take. It was like going from living in a mansion to moving into a cardboard box in an alley.

We stepped across to the back deck with our gear, walked through the large shark cage and got acquainted with the rest of the crew. Mike was another volunteer who had a passion for sharks, and a smiling Steve was the guy who had supplied the boat with all of the fish carcasses in the coolers. Steve was smiling because he had already laid claim to the largest and only sleeping quarters on the boat besides the captain's cabin. Carrie and I, being the last to arrive, got to choose between a stinky couch and the carpeted deck. If Carrie had been just a little bit stronger, she would have gotten the couch.

The high-tech dive boat pulled away and headed for the island. I watched it cut through the swell towards the lee side of Santa Barbara Island through a dirty port hole, streaked with bird droppings and rust. The faint odor of diesel exhaust inside the cabin and my stinky frat house couch dropped my mood considerably. I really wanted to be on the other boat.

Carrie and I stowed our gear, and then we climbed to the upper deck to introduce ourselves to the captain. He seemed even less interested in our face-to-face greeting, and again rudely dismissed us to work on something. We returned to the main cabin and decided that we may as well start the chum slick. Steve walked over to the stinky coolers and started to remove the duct tape keeping the lids close. He had somehow obtained four large coolers worth of fish carcasses. He had been storing the coolers in a meat locker and had picked them up on his way to the dock earlier that day. To keep any putrid juices from spilling out, he had closed the lids and duct taped them shut around the seam. Now, eight hours later, we shouldn't have been surprised to open them up and find that they had been effectively acting like coolers. The solid, immovable block of icy ooze inside weighed close to 250 pounds. Chumming would have to wait.

Mike and I grabbed one of the coolers and lifted it over the back of the boat and floated it in the water to thaw. We lifted the lids on the other three containers to expose the insides to the mid-day sun. The conditions couldn't have been worse for chumming. The ocean was flat calm and there was absolutely no wind. To top things off, the majority of the fish in the coolers were rockfish carcasses that didn't possess a great deal of fish oil. To lay out a good chum slick, it's best to use the oiliest fish you can find. While chunks of floating fish will attract sharks nearby, fish oil will hang near the surface and spread the scent out for miles. A water current or moving water through slightly windy conditions ensure that the chum slick will travel great distances, increasing our odds of attracting sharks. A shark swimming miles away will detect the slick and follow it right to the boat. With no wind, no current and very little fish oil, the only thing we'd be attracting is sea gulls.

I walked over to Carrie, who was sitting on the back deck smoking a cigarette. "You know this isn't going to work, right?" I said. She just shrugged and flicked ashes into the flat calm waters. I walked over to the stern and looked down at the cooler floating at the back of the boat. The block of ice inside was slowly melting, but in the present conditions it wasn't going to do us much good. At this point we weren't really a vessel laying down a chum slick; we were more a boat just leaking fish ooze.

Since there wasn't much to do but wait, everyone left the back deck and found other things to occupy their time. Steve retired to the master suite and I didn't see him again until we pulled into the harbor two days later. Carrie grabbed a book and a cigarette and parked herself inside the shark cage on the back deck. The captain made sure that the upper deck chairs didn't float away by lying in both of them and catching an afternoon nap. I grabbed a fishing rod and decided to see what was biting on the bottom.

I walked out to the back deck, rummaged through my fishing gear and found a fairly heavy silver jig. I flipped the reel release and lightly tossed the jig a few yards from the stern. I watched the line peel off the reel and applied enough thumb pressure to keep the line from lashing back on itself. We were anchored in 230 feet of water and the bright shiny lure had no problem finding the bottom in the flat conditions. I flipped the reel back in gear and jigged the lure off the bottom three times before something grabbed it. I set the hook and began reeling the fish to the surface. The bright-orange starry rockfish hardly struggled as it came to the boat. Since the chum was still a solid mass of ice, I decided to save whatever I caught so we could start a chum slick with the fresh fish. Besides, after popping up from over seven atmospheres, the fish I started catching were in no shape for release. The change in pressure from their home on the bottom to the back deck of the stinky boat caused their eyes to pop out of their sockets and their stomachs to protrude, inverted from their mouths. Not a pleasant end to the trip.

In about an hour, I had filled a five-gallon bucket with over a dozen starry rockfish. The captain came down to the back deck dragging a contraption and dropped it heavily at his feet. The old garbage disposal was bolted with heavy hardware to an old milk crate. The captain picked the unit up and placed it on a cooler near the stern. He unwound several feet of dirty blue hose and tossed it over the side. He grabbed a screwdriver and tightened a few screws on the unit. He placed the screwdriver in his mouth and shook the unit with both hands for some reason. I watched as he tossed the electrical cord towards an outlet near the stern and then grabbed the blue hose. With the screwdriver still in his mouth, he turned towards me. "Dosth spart dish undull dis toast distober da zide," he mumbled, shaking the hose for emphasis. He then turned without waiting for a response. I assumed the hose where the fish gunk came out was to be positioned over the side before we started the machine. I could tell the captain wasn't much of a talker. In fact, those garbled instructions were the only words he spoke to me over the entire trip.

The captain positioned the flesh-grinder near the back of the boat and plugged it in. He reached down and grabbed my bucket of fish and dropped it roughly near the cooler holding the grinder. He flicked the large switch at the top of the disposal and the contraption came to life with an angry, whirling growl. He grabbed one of the orange fish from the bucket and dropped it into the top of the disposal. The captain then shoved the fish into the opening with a short piece of broom stick. The disposal twirled the fish, caught the head, grinded loudly for several seconds and then died. "Bruthafrugin pizzachit." He flipped the switch off, and then on again and a small spark jumped from the unit. After a few attempts at repair, the machine was unplugged, wrapped with its own dirty hose and placed in the corner. The electric fish grinder was done.

The cooler we had hung over the side earlier in the day was actually thawing nicely. There was still not much of a current, but the pieces were being freed up from the frozen block and drifting into the deep. Small bait fish were slipping in from under the boat and grabbing what their little mouths would allow. I grabbed the rope tied to the cooler and tugged on it sharply. The movement caused a thick cloud of fish bits to burst from the cooler and drift towards the bottom. With the present ocean conditions, there really wasn't much else we could do. With food in the water and the others coolers thawing, the chum boat was chumming despite the conditions.

At a little past noon, the dive boat ran out and dropped off lunches and dinners for the chumming crew. As we ate, we discussed how we could maximize our efforts with the present bait and ocean conditions. The captain had set up a hand-cranked grinder at the stern for grinding fish when the high-tech unit died. With what I had caught earlier and the thawing deck coolers, we had a dozen gallons of bloody ooze sitting in a large plastic drum near the shark cage. We decided to start chumming in shifts, each of us scooping out slurry over a three-hour period.

I walked out to the back deck after dinner and looked around. The water was still flat calm, but I could see the chum slick moving away at more of an angle indicating a current. I spent the last hour of daylight fishing and adding more fish to the grind bucket. I watched the horizon swallow up the blurry sun until it was nothing more than an orange smudge at the edge of the ocean. The nearby island, once rocky and proud, now without the sun looked like a wide tear in the sky. The catamaran sat tucked back into one of the bays and I could see people on the back deck. They looked to be having a small party in the shadow of the island. A wave of envy shot through me. I wondered what the hell I was doing out there.

I grabbed the fishing rod and sent the heavy jig back to the bottom. I clicked the reel release, reeled in the slack and started snapping the rod up and down. Not ten seconds later, something larger grabbed the lure on the ocean floor. I set the hook on instinct and began quickly reeling the fish to the surface.

I supposed it was another starry rockfish. The fish that popped to the surface wasn't orange. I lifted it aboard and caught the lure in my hand and held the fish up for inspection. I could tell it was part of the rockfish family, but it didn't look like anything I had ever seen. During my fisheries training I had taken great pride in being able to identify any fish off our coast. That included the more than fifty species of rockfish that frequented the depths. However, holding the fish dangling from the heavy spoon in the dim beam of the deck light, I had no idea what it was.

I unhooked the fish, placed it in a bucket and set it behind the old useless electrical grinder so it wouldn't get accidentally grinded up. I went inside and leafed through the two offshore fish identification books I had brought along on the trip. After several minutes of using the book keys, I was absolutely sure the specimen in the bucket was part of the rockfish family, but still I couldn't figure out which species it was. I walked out to the back deck and examined the fish for a few more minutes and then tucked the bucket in the corner.

I walked back into the cabin and retired to my stinky couch. We were starting our rotational chumming deal tonight and for some reason I was scheduled for a shift early the next morning. As I tucked myself into my sleeping bag, Mike stopped by. "You're on for 5:00 AM, right Tim?"

he asked, passing me on his way to the chum bucket. I gave him some mumbled response that neither confirmed, nor denied his statement. As I waited for sleep, I thought about the new species sitting in the bucket on the back deck. Even though you weren't supposed to name a newly discovered species after yourself, I couldn't help but to try out a few names. I imagined a full-color spread of the new species in *National Geographic*, with me on the cover. I smiled and closed my eyes.

The next morning Mike woke me up while it was still dark. He was standing over me with a stinky fish ladle and asked if I was still asleep. What a completely useless comment if you think about it. I eased off the couch and rubbed my eyes. He mentioned that it was getting a little rough out there. Sure enough, I could tell that the boat was swinging on the anchor and moving a bit more as a swell developed. Mike thrust the chum ladle into my hand and turned before I could swing it at him. I hung my throbbing head in my hands. I couldn't shake the pounding headache I had woken up with. I also didn't feel at all like standing on the back of this barge in heavy seas ladling fish scum. As I wrestled myself from sleep, a large drop of fish ooze dropped from the ladle and landed on my sock. Just before I slipped on my shoe, I wiped the slime on the couch.

I stumbled out to the back of the boat. The wind was blowing and the water's surface was choppy. The slight swell made it difficult to maintain balance while I tended the chum slick and just standing out on the back deck, I was getting wet. I leaned on the transom and peered into the depths, the lone deck light my only source of illumination. As my eyes adjusted, I could see something in the chum slick. I shaded my eyes and a blue rocket shot through the available light and then out of sight on the portside. The shark turned quickly in the darkness, swung back around and swam close to the back of the boat through the slick. We didn't have any large chunks of meat to hang off the back of the boat to keep larger sharks interested in sticking around. All the fish carcasses in the cooler were about twelve inches long and I hadn't caught anything larger from the back of the boat. Consequently, if large sharks came into the slick, they wouldn't find much to keep them hanging around.

The five-foot mako circled once in the slick and then headed out towards the open ocean faster than my eyes could follow. It was gone.

I stared at the water and continued to shuffle to keep my balance. Grabbing the heavy bars of the shark tank to steady myself, I wondered what the hell I was doing out there. I glanced back towards the island. The dive boat was nestled comfortably in the back of a protected cove and the only light beaming from the ship was a rear deck light. I had a hard time believing that anyone would brave the hour and the conditions to jump in the dark, cold water with a shark.

I grabbed a five-gallon bucket and dipped it into the large chum reservoir. I let the ooze fill half the bucket and I heaved the now bloody container to the edge of the boat and poured the entire thing over the side. The water in the light turned dark and white pieces of fish floated away with the current. I waited another ten minutes and repeated the bucket chum. I rinsed the bucket off at the back of the boat and tossed it in the corner. I walked back inside, climbed into my sleeping bag and fell asleep on the couch.

Sunlight through the side window woke me from a dream about tractors. I sat up with another pounding headache and cradled my throbbing head in my hands. I glanced out to the back deck. Carrie was manning the hand-grinder and smoking a cigarette. Her image was fogged and distorted by the diesel fumes floating through the floor boards of the boat, filling our sleeping quarters with poisonous gas. Fumes from the generator were filling the cabin.

I stood up blinking my eyes. The mystery behind my tractor dreams and my constant headaches was solved. At this point fresh air was not just refreshing, but essential. I stumbled out to the back deck, thankful that by the end of the day I would be sleeping in my own, gas free bed. As I stood next to Carrie, a waft of cigarette smoke engulfed my head and it was almost refreshing.

"How did it go this morning?" Carrie asked.

"It wasn't too bad," I said. My guilt on cutting my morning short dissipated completely as I saw what she was doing. Carrie had been grinding fish from the same bucket I had used to store my priceless specimen. I grabbed the bucket and looked inside. Two bright orange fish and nothing else made up the contents of the bucket. "Where..." I said.

"What's your problem?" she asked.

It was pretty clear that storing the special fish anywhere close to the teeth of the grinder was a huge mistake. "Carrie, there was a different species of fish in here besides the..."

Carrie stopped what she was doing and looked at me. "Oh yeah, I put it in the grinder," she said matter-of-factly.

I closed my eyes. My histrionics were lost on Carrie. The fish had been dredged up from the depths to fulfill its destiny with science, only to be returned to the bottom a few hours later in pieces. I leaned over the side and saw large white fish chunks drifting into the darkness. My fish, whatever it was, was gone.

Shortly after the sun came up, the dive conditions at the island began to deteriorate. The increasing swell had found the island shore, stirring up visibility beyond what the average diver found enjoyable. The deck hand on the chum boat informed us that the catamaran was running out to pick up any crew members that wanted to head in with them. Carrie and I carefully weighed our options on returning with the divers on the nice boat or staying in the floating gas chamber. The decision was an easy one.

An hour later, the catamaran again nosed up to the back of the chum boat, easing up and hesitating to get close. Carrie and I transferred our gear to the other boat and jumped back to luxury. We enjoyed a gourmet lunch back to the harbor, and breathed fresh air that wouldn't eventually kill us. During the trip back, Carrie and I gave a slide presentation on our local shark species to the paying divers and conversed with some of the patrons as we ran. Most appreciated the effort we had made on the chum boat, and a few appeared a little relieved that no white sharks had been sighted.

The clear picture I have in my mind of the mystery fish remains. The several searches I have made since the trip regarding this species have been fruitless. No fish currently described fits the description of what I caught that day, and I frequently kick myself for not taking a photo of the fish. I'm sure that with time the image will slowly fade from my memory as will the pains of the opportunity lost. What I won't forget is the image of Carrie running the hand grinder through the fog of diesel fumes and the sight of the white fish chunks drifting slowly to the bottom of the ocean.

STRANDED

During my last summer as a graduate student I accepted a job as a field operations manager for the fisheries program. The position involved planning and participating in sampling trips that involved collecting specimens for university projects. In reality, I was already doing just that, but now I not only received an official title, I'd be getting paid a bit more for it as well.

While we were finishing up with our spotted sand bass research in the lab, several other projects we were involved with were still in the sample collecting phase. We had made two sampling trips to Mexico to collect gold spotted bass. The one in Oke Landing had been successful. Another a few months later in La Paz had not. To boost the sample size and to expand the sample populations, Larry planned a trip to the Bay of Los Angeles, a small seaside town about half way down the Baja Peninsula. Once again sightings of the golden fish were reported from the area and that information was good enough to set the traveling fish lab into action.

A group of students and scientist were once again traveling the straight road into Mexico about three weeks later. Our caravan pulled into the sleepy seaside town of Bay of Los Angeles late in the afternoon. We set up camp at a beach side spot that had the room to accommodate our large group and was complete with restrooms and primitive showers. At the shore, the warm gulf water scarcely moved as the sea slowly undulated near the sand. Out towards the horizon in the fading light, the dark brown Angel de La Guardia islands sat teasingly close several miles off shore. They sat like rocky guardians at the edge of the huge bay. Even though they looked anything but inviting, I was anxious to get out there and explore both above and below the water.

That evening, around the fire, Larry handed out boat and sampling assignments for the following day. We had towed down three small skiffs with us and loading them with people that could fish was really all that was required for the sampling. Larry wanted at least 250 gold spotted bass specimens from this location to equal our bounty from Oke Landing. As far as fishing experience, as long as you could turn the handle on the reel, you qualified for catching gold spotted bass. Dropping a shiny lure laced with squid to the bottom in 250 feet of water was really all you needed to do to get a bite.

The next two days were as perfect as conditions and sample fishing could be. The Gulf treated us to flat conditions and once we found the school of fish, the bite was wide open. Every drop of the lure resulted in a valued specimen for the study coming over the rail and finding its final resting place in a crowded five-gallon bucket. Illustrating the chaos to acquire food at the bottom, some lures came back to the surface with two gold spotted bass hanging from the treble hooks. With the excellent conditions and the incredible fishing, we had all the samples we needed by the morning of the second day. With the efficiency that comes with mastering a series of mundane tasks, and the large number of volunteers, we finished working through all of the specimens just before dinner.

That evening we celebrated our success and made plans to explore the offshore islands the following morning. In the flash of the dimming embers of the camp fire, we relaxed and felt the warmth of accomplishment and the end of a very good day in Mexico.

On the afternoon of day three, I stepped from my tent and looked out towards the islands. The weather had turned ugly. The calm, smooth conditions that had greeted us earlier that morning had now given way to a truly angry sea. Our group had been out there earlier that day when the Gulf was calm, serene and happy. Now it looked about as inviting as used bath water in a blender. And I knew I had to go back out there.

Earlier in the day, several of the group had asked if I'd ferry them out to one of the islands so they could do some diving. Part of my position involved running the boats in pursuit of science. This didn't mean we couldn't have some fun during the down time and I didn't think twice about taking them out.

That morning we were out on the water at sunrise. Two of the skiffs were loaded with dive gear and fishing rods and headed for the distant islands loaded for fun. I was transporting the four divers, with all their gear to a sandy beach that offered easy access for me and an easy dive entry for them. Greg was piloting the other boat full of fishermen to an underwater drop off a few miles off the coast of the same island.

The water was like glass and we made the eight-mile run to the islands in no time. I peeled off from Greg with a wave and headed to the beach to drop off the divers. Greg returned the gesture and headed for the unseen canyon in the gulf. I eased the bow towards the shore and cut the engine as we approached. Curtis jumped out and held the boat in place as his group offloaded their gear. The waves were non-existent and in seconds I was the only one on board. I handed Curtis a radio and told him I'd be back to check on them in a few hours. We did a quick radio check and he pushed the boat out. I trimmed the engine up, started it and slowly backed away from shore. An easier offload I could not remember.

I looked out towards the flat horizon and found the other boat a few miles out, floating near the drop off. I eased the throttle forward and pointed the bow towards the tiny boat. I covered the distance in a minute. I drifted up next to the other Whaler and picked up John to spread out the anglers. We motored a short distance away and I shut down the engine. I grabbed a boat rod and sent the heavy silver lure, tipped with a whole squid, to the ocean floor 250 feet below. Even though we had fished hard for science for the last two days, fishing for fun was the first thing we thought of when we were allowed free time.

One of the things we had been warned about when we first started coming to the Gulf of California, were the high offshore winds that would instantly appear. The Chubascos, as the locals called them, would turn the serene waters into a dangerous froth. The sudden winds would whip the flat conditions into a frenzy of large, boat-swallowing waves that had no pattern and came from all directions. The stories of huge seas shutting down the gulf for days at a time were not only legendary but well documented. Guide books would warn travelers that if even a wisp of heavy wind was detected, they need to quickly make their way to the safety of the shore before things got deadly.

We had just finished our second drift, when a strong and sudden wind gust caught us totally off guard. The wind turned both boats and started to push them out away from the islands. It had gone from flat calm to four-foot breakers in less than two minutes. The horizon was now occupied with developing white caps and we were being pushed so fast that I knew we were done fishing for the day. We were also taking enough wind spray and water over the bow for me to become a bit concerned.

By the time we stowed the gear and got ready to head back, the swell had grown to six feet and we were still eight miles away from the safety of the harbor. The waves were coming from one general direction, but the sea was chaotic and navigating back in the 18 foot skiffs was going to be slow and dangerous. I looked over to Greg, who had just started his boat and motioned him over. I could see the heavy concern on his face and I knew that he had little experience maneuvering a small boat in these conditions. Staying calm on my part and his would be the key to getting back safely.

Greg's boat had two other people on board besides himself. The buckets we had brought along to store fish in were now going to be used for bailing water out of the skiffs. As Greg pulled close, I could see the fishermen stowing the gear and clearing the deck for the ride back. I looked over and gave him a sturdy nod and a look that simply stated that we can do this. I told him to take it slow and keep the nose of the boat pointed into the swell. Greg looked over at me and I could see he was genuinely scared.

I pointed our Whaler towards the beach where I had dropped the divers and motioned for him to follow me. It would take us 40 minutes to travel the two miles to the cove.

Early attempts to reach Curtis on the radio were useless and we were out too far to see if they were still on the beach. It had been over an hour since I had dropped them off and I hoped they were still on shore. I watched the violent sea rise and fall all around us. I tightened my grip on the steering wheel. For a few seconds, I seriously wondered what I was going to do. Then I realized it wasn't up to me. The answer to the situation was all around me. There's no way I was going to safely pick up the dive group in the degraded conditions.

I kept the boat in gear as I came into the cove. I had to roughly slam the engine from forward to reverse several times just to fight the waves that wanted to push me up onto the beach. I could see the divers at the shore very close to where I had dropped them off. I felt the knot in my stomach tighten. I was not looking forward to telling them that their island adventure would have to be extended, indefinitely.

Before I got too close, I turned around to check on Greg. He had wisely stayed out away from shore to wait for me. I eased in as close as the degrading conditions would allow and motioned to Curtis to meet me down the beach. After several minutes of struggling we were finally able to establish a high technology radio link between the 75-yards that separated us.

I took a deep breath. "Curtis," I started, "I can't pick you up right now it's just too rough." After several seconds of static he responded. "No problem, Tim," he said rather matter-of-factly. I could tell he really wasn't grasping the seriousness of our situation. I told him I may not be able to get back out here until just before dark. He stared at the radio and then spoke into it "Don't worry, we'll be O.K." I could hear the thundering smash of the shore break in his transmission and I knew right now I could not get any closer. I instructed Curtis to monitor the same channel and not to leave the radio on and run down the battery. He confirmed that he would try every half hour to reach me. I also made sure they had snacks and plenty of water.

I waved to him and he waved back. As we turned to head back out, I saw him jog to the group completely unconcerned with the situation. I had to admire his attitude.

I motored out towards Greg and we started out of the cove in single file. Just as we were leaving, Curtis' excited voice came back over the radio. "Hey Tim, don't worry about us, we came pr–," and then he broke off. I turned back towards the shore and saw Curtis waving and holding something about ten inches long, shaped like a carrot and purple in his hand. "What the hell is that?" I mumbled. John looked over. "Is that what I think it is?" John muttered. I didn't know what dirty little path John's demented mind was traveling, but I found myself on the same road. As Curtis turned, tossing and catching the "toy", I suddenly realized that Curtis was the only male in the group stranded on the island. John and I both looked at each other and started laughing. It suddenly became clear to us why Curtis wasn't too upset at his extended stay on the isolated beach.

I motored up to Greg's vessel and we headed out of the cove. As we rounded the corner we were met head-on by the worst sea conditions imaginable. In the short time I had been talking to Curtis, the rolling five and six-foot swells had graduated to eight footers, with the occasional larger rogue waved added into the mix. I quickly glanced back towards Greg and saw that fear was permanently etched on his face. As I grabbed the steering wheel, I noticed my own hands were shaking.

A slap on the back from John brought me back to camp. "That was scary as hell," he said taking a swig of his beer. It had taken us over two hours to return to the safety of the harbor. We had dodged eight and ten foot swells and monstrous waves. John had spent the entire time bailing water out of our boat and Greg had handled and maneuvered his vessel like a seasoned veteran. Three hours had passed since our return and the sea had shown no signs of calming down. I calculated how long it had taken us to get back and I knew that if I didn't go out and pick up the stranded divers soon, I would be traveling back in darkness. "I have to go back out there now and get them," I said.

John nodded. "I'll hook up the boat."

Within minutes John pulled up to camp towing the Whaler. For some reason, the boat looked so incredibly small and a wave of true fear washed through as I realized what I was about to do. Since I had to pick up four people and their gear in rough conditions, I needed all the room I could get. I jumped into the Whaler and removed everything I didn't need and made sure all was in order. I knew as a last resort, I could have the group leave their gear on the island and we could pick it up at a later time when it was calmer, whenever that would be.

I put two full gas cans on board, and checked the one attached to the engine. I knelt down and pulled on the boat plug, and then tested the steering linkage and prop on the motor. I grabbed the radio sitting on the center console and checked the battery. It was done. I knew the battery had died during Curtis' last transmission. There was no way I was leaving the shore without a working radio. Danny and Larry searched the camp and found a working unit. Danny handed me an older radio that had been manufactured before I was born. "This one works" he said, "just don't get it wet."

John jumped into the cab of his truck and looked back to the Whaler. After checking a few more things, I nodded and pointed towards the harbor. Easing out of camp I saw Larry standing at the front of the truck looking my way. I had been running boats for Larry for years and we both knew I had never been up against seas like this before, not in a small boat. And because I had a full load to pick up, I had to go alone. He looked me straight in the eye, nodded and raised his beer towards me. No words needed to be exchanged. I knew what he was saying and I returned the nod.

I sat bouncing in the boat on the short drive to the ramp. I looked out towards the islands and it looked ugly. The sky was gray and the rocky crags out on the horizon looked like the forbidden land in a scary movie. If you had a choice, not a single cell in your body would head out towards them, but I had no choice.

At the harbor, John maneuvered the Whaler back to the ramp and backed me down into the water. I started the engine, tossed a bail bucket into the bow and started backing off the trailer. "Hey," John yelled. I looked up and John tossed me a cold beer. I cracked the cap and downed the entire thing. John gave me an approving grin as I tossed him the empty. "Hey," I said, "when I get back, I'll buzz by the camp." He nodded and I slowly backed the small boat off the trailer. As I did, John jumped up onto the tailgate of his truck and yelled out one of his heavy-metal mantras. I considered myself blessed and turned the vessel towards the mouth of the harbor. I placed the antique radio on the console and prepared for the worst.

For the first few miles the going was easy over the six and eight-foot swells in the large, somewhat protected harbor. Just as I was feeling like things weren't going too badly, a ten foot rogue wave hit the front of the Whaler. For the briefest of moments the nose of the boat was gone, buried in the crest. The wave jarred the boat severely and the outboard sputtered, hesitated and then caught itself. Water rolled from the front of the skiff to the back with such force that when it hit the transom, the backsplash completely soaked me. I squeezed the wheel tighter and physically tried to shake the fear from me. I knew if the boat died out here, I was in real trouble.

I shook off the panic and decided to see if I could contact Curtis with the radio. I reached towards the console and the panic returned. The radio was gone. I quickly searched the deck and checked the bilge. There, laying in several inches of water was Danny's relic. I picked it up and several ounces of water dripped from within. I dried off the useless instrument and shoved it in my pocket, quite surprised when I found that it fit. I was now truly on my own.

I spent the next hour and a half dodging eight and ten-foot waves, attempting to bail out the boat as I motored. The deck of the skiff tipped side to side at drastic angles constantly. My legs were weary and my knees were bruised and sore from constantly smacking the center console. I had taken several waves over the bow and was now completely drenched. I lost my hat somewhere in the channel and the bailing bucket was now resting comfortably at the bottom of the ocean. As I littered the sea floor with gear, I remember having to seriously push through the thought of turning around and heading back to land and safety.

I felt myself chopping up the trip into attainable sections to not only keep my mind focused, but to measure my progress. I had already made it out of the harbor, which wasn't monumental, but it marked the first leg of my voyage. The next goal was to reach the narrower channel between the two larger islands. I was hopeful that once I made it there I may experience some calmer waters in the lee. I could then make up some time running to the dive island, a smaller rocky crag that sat another three miles further out. Unfortunately, as I floated into the channel, I noticed that the swell was coming between the two islands and it offered me no protection. However, I did feel a bit calmer being closer to the desolate rocky islands in the channel. I figured if something went very wrong, I could at least make it to the shore of one of the islands.

Close to an hour after cutting through the two islands, I reached the front side of the dive island. As I rounded the corner of the cove, I noticed that the group on shore had built a rather large fire. A few yards away, a large pile of gear sat close to the angry shore in preparation for the pickup. Entering the cove, the waves dropped in size considerably. Conditions that would have stressed me on a calm day were now a very welcome sight in the small, protected cove. I felt a substantial drop in tension as I maneuvered the smaller waves, less than half the size of the monsters out in the main channel that were trying to kill me.

I motored up as close to the shore as I dared and again had to worked the throttle roughly. The group saw me approach and I could tell they were glad to see me. Curtis ran down to the shore holding up the radio. I just shook my head and held out my hands indicating that that wasn't going to work. He nodded that he understood and he waited for instructions. I waited for a slight lull in the pounding shore break noise. Yelling loudly, I told him that they may need to get wet and we have to do this quickly. He waved and then headed back to the group so they could get their gear. I pointed towards a small group of boulders that were set far inside the cove and appeared to be somewhat protected. The group met me there and I eased the small boat in close between the swell. Curtis tossed in the dive gear first. Half steering, I moved the gear way from the bow so the group could board safely. In a matter of seconds they were all aboard and I eased the boat away from the rocky shore. We distributed all the weight on the boat as evenly as we could. Almost all the gear was directly in front of the center console and all four passengers, two on each side, were located behind it. I turned the boat towards the mouth of the cove and we headed for home. It wasn't quite 6:00 PM.

As we approached the corner, I told everyone to stay low and hold on. "It's going to be a little rough outside," I said. Cindy, a volunteer for this trip, looked up and asked me if it was going to be rougher than it was in the cove. I just looked over to her and smiled. As we turned the corner and hit the channel and the huge waves, I heard her mutter, "Oh my God!"

I felt just slightly relieved that I had the group in the boat and headed for home. However, I was completely aware that I still had a monstrous task ahead of me and daylight was fading quickly. I squeezed the steering wheel until my hands hurt and I sucked in a deep, soul-cleansing breath and let it out. I re-stacked my determination and grabbed the throttle with my right hand. It was time to get it done.

We traveled up the crest and down into the troughs of hundreds of eight and ten-foot waves. We were surrounded by white caps and water regularly peaked over the bow and sloshed to the bilge. I had Curtis man the boat plug, pulling it out periodically to let the water race out of the Whaler. Without a bailing bucket, it was all we could do to keep the water level inside the boat to a minimum. The wind was harsh and seemed to come from all directions. I did my best to keep us away from the rogue waves and tried to keep my mind from thinking how ridiculously close to the edge we were. As we came to the top of one wave, the bow remained skyward for what seemed like minutes and it almost looked like we were flying. When it came back down, the hull slammed into the trough and drenched everyone. Once again the motor sputtered, hesitated and caught itself. I tightened my grip on the steering wheel and took a nervous breath. I glanced back to the engine to make sure it was still there.

Water continuously sloshed over the sides of the whaler. Everyone was drenched and several inches of water washed back and forth on the deck. Miles from the safety of the harbor, I remember having a fleeting thought that we weren't going to make it. Just then I felt someone tugging on my shirt. I looked down and saw Curtis kneeling at the side of the boat with a huge grin on his face. He gave me the thumbs up sign and he actually looked like he was enjoying the ride. You really had to admire his attitude. I then started to wonder what had happened out on that island.

After we left the cove, we stayed in the center of the channel between the two larger islands. The water was chaotic and there was no pattern. Waves came from both shores and slammed together all around us. It was like trying to navigate in washing machine.

By my estimation we were still about five miles from the safety of the harbor. The shore would appear and then disappear as we dropped into the troughs; the walls of water towering over the boat threatening to swallow us. My forearm muscles began to ache, the result of my lengthy death grip on the steering wheel. My legs were locked and even though I was soaked to the bone, my mouth was as dry as desert sand. From the position of the islands, I could tell we were making slow progress, but this did little to ease my mind. I wanted this crossing over with and was getting tired of having to be so alert constantly. To their credit, the dive group remained calm and cooperative.

We had just taken another medium sized wave over the front that drenched the starboard side occupants, when I felt another tug on my shirt. I turned and looked down to see Cindy with her left hand cupped over one of her eyes. At first I thought she was injured and I leaned in close to see what had happened. She looked up at me with her one eye and spoke. "My contact lens just fell out!" My mind flashed to the transparent lens sinking to the muddy bottom of the gulf hundreds of feet below. I briefly smiled as I thought that losing a contact lens out here was like walking across the country and misplacing your car keys somewhere along the way. It was just gone.

The angry sea showed absolutely no sign of relenting and the occupants, as well as the captain were getting weary. The lights on the now dark shore danced as we were rocked and rolled over the swells and between the rogues. I pointed the beaten craft towards the largest of the lights; a large campfire that I was convinced was the beacon from our camp. The sun had completely set and in a matter of minutes forging this body of water in its current conditions would become exponentially more dangerous. I tightened my grip on the stainless steel steering wheel and adjusted my stance. We needed to get home and soon.

Approximately two miles from the harbor the conditions began to improve. The swell had dropped to about five feet and it looked like we had left the rogue waves in the channel. Even though I couldn't see it, I knew the calmer water meant we were close to the harbor. The lights on shore were not as animated and the large fire in camp could be clearly seen. The outline of the mainland was barely visible in the fading light and I had to slow down and think where the actual harbor was located in the shadows of the shore. Confident that I could locate the safety of the harbor from our camp, I angled the Whaler towards the beacon and breathed a little easier. As I shook the weariness from my body, I wondered how long my legs had been shaking.

About twenty minutes after sunset, we were racing by the camp hooting and hollering. The gang on shore cheered as we motored by and I saw John jump from his lawn chair and head towards his truck to meet us at the ramp. We had survived!

To this day, I can't think of a sweeter feeling than when that boat nestled itself onto the trailer back at the harbor. John jumped from his truck, clipped the boat to the drag belt on the trailer and started winding us in snug. "I'll bet that was a hell of a ride," he said, smiling. I looked over at him, relaxing for the first time in five hours. "Never again," I said.

Once back at camp, Cheryl met me with a cold beer and a carne asada burrito. Food and drink had never tasted so good. Larry approached the boat. "Good job," he said, giving me a fatherly hug. "I was getting a little worried, we hadn't heard from you," he added. I reached into my soaking pocket and pulled out the old radio that was now nothing more than a paper weight. "Do me a favor," I said, "give this back to Danny." I handed Larry the radio. "Make sure I'm not around when you do," I added. Larry turned the now useless instrument over in his hand. "Did you get it wet?" He asked. There wasn't a spot on me that was dry, salt crystals had formed in my hair and I sloshed when I walked. "No", I said, laughing.

After a quick change of clothes, I approached the fire to warm up and relax. As I sat putting on dry socks, I listened as Curtis explained how they had survived the ordeal. I was just about to ask him how they had started the fire on shore, when he produced the ten-inch purple instrument. John and I instantly glanced at each other. Who knows why Cindy brought it out with her on an island dive and knowing her, you really wouldn't think she'd have a need for such a device, but here it was. The ten-inch, butane curling iron apparently was instrumental in the starting of the fire they had on shore. John and I went into hysterics. I was still wiping tears from my eyes when I looked across the fire and saw Danny take the battery panel off the back of his old radio. Several ounces of seawater dribbled out of the compartment, onto his shirt. "TIM!" he yelled. I don't really remember what happened after that; all I know is that I was laughing too hard to run.

SEAL BEACH

Towards the end of that last graduate summer, I got the opportunity to work on another fisheries contract for the university. Since I had completed the course requirements for my advanced degree, I wanted to spend more time in the fisheries field, and move away from teaching lab classes.

As soon as I had entered graduate school I was able to supplement my meager paycheck by teaching entry-level biology labs. While the money for teaching on campus was good, I had realized early on that molding young minds was not for me. We often deemed the endeavor thirteenth grade, having to deal with mass family deaths that all seem to occur a week prior to finals. A cagey attempt by me to capture a pair of known cheaters during a weekly quiz resulted in half the class being swept up in the dragnet. I sadly realized that most of my best students were just really good at deceit.

During my last teaching semester, one of my best students came to me a week before the final. She said that she would be unable to attend the test because she had a brain tumor and was scheduled for surgery the following afternoon. I was tempted to have her put the unsightly growth in a jar and bring it in to confirm her story. However, since she had aced all her previous quizzes, I gave her the grade she had earned thus far in the class. About a month later I saw her in the hallway of the science building. She had a shaved head, and what looked like a C-shaped scar on the side of her skull. She walked right on by me as if she had never seen me before. Now that's dedication.

The last straw came during my final semester. A near mutiny after a lab practical took me completely by surprise. The class had spent the last hour weaving their way through the set microscopes and lab specimens with numbered flags in their guts, identifying different structures of the rubbery animals. When they returned after a break, they made it very clear that the test was too hard for their smooth brains. The complaining started at one end of the class, and like a tidal wave washed over the rest of the group until all were loudly voicing their illiterate objections. Primitively pushed into a corner, I responded to the verbal havoc with an even louder, expletive-laden retort. I remember feeling like a foul-mouthed demon had invaded my body, and together we floated over the drooling idiots, spraying them with nasty invectives and fevered spit. The F-word flew from my mouth as noun, verb and adjective. When I was done, the class was quiet, timid and cowering. As they filed out, a few apologized for the complaining. On that day, my teaching career came to an end and when the opportunity came to leave teaching behind, I gladly took it.

The fisheries contract was one I had worked on before. The California Department of Fish and Game had enlisted the university field staff to survey and collect white sea bass off the coast to assess an enhancement program. For several years the State had been involved in an elaborate project to replenish the dwindling white sea bass populations through hatchery enhancement. A hatchery facility in Carlsbad, California supplied fingerlings to a number of grow-out pens along the southern California coast. These pens were manned by volunteers who fed the precious stock fish pellets on a daily basis. Once the white sea bass reached a hearty and one hoped survivable size, the pen gates were opened up and the fish, hand-held by man to this point, were released to fend for themselves. After spending their developmental months in tanks and grow-out pens, feeding on drab pellets, one had to wonder how they would fare out in the unforgiving wild ocean. I would imagine come feeding time, most would stare in frothy anticipation to the surface waiting for the crap-colored pellets of life to fall from the wet heavens.

Before the fingerlings were moved to the grow out pens, each fish had been affixed at the hatchery with a small wire tag injected into the fish's cheek. When the fish was recaptured, these coded tags could then be read by a special magnetic reader we called the wand. The university had been contracted to set gillnets up and down the coast in the hopes of encountering the released fish, thus assessing the project's relative success over time. Since I had worked this particular contract before, I was recently promoted to lead the sampling team during our gillnet surveys. Every other month our survey group would head out on the *Yellowfin* and set the nets at randomly chosen areas along the coast overnight. The next morning, we'd collect the nets, work up the catch and head for another sample spot. During the week-long trips, we'd usually hit four or five sample locations during a survey trip.

There was no denying that white sea bass numbers off the California coast had increased substantially since the early 1990's. Coinciding with mass releases of pen-raised fish from the hatchery, it was hard not to see the program as a success. However, large, healthy looking adult fish that were being caught in near historical numbers lacked the one thing that would ring the resounding bell of success for the hatchery enhancement program: none of the big fish had the cheek tag that would declare these white sea bass as having come from the other side of the fish tracks. While the public supported the hatchery program, fisheries scientists saw the natural increase in the white sea bass population as an outcome of the banning of near-shore commercial gill netting in the early 90s.

Research had shown that the near-shore environment was the nursery grounds for many marine fish species. Relentless and greedy gillnetting for commercial gain in these areas had retarded growth prospects. Large, ready-to-spawn adults would come into the shallows to spawn only to be captured and killed in the nets. Once the ban limited the distance from shore a commercial fishermen could deploy his gear, the nursery grounds were once again open, and the numbers increased naturally.

The irony of using the same gear to assess the hatchery program that had been responsible for stunting the white sea bass population, was never lost on me during these surveys. As sampling gear goes, there was nothing more efficient and deadly than gillnetting. However, no matter what the project or its eventual goal, I could never quite come to grips with the enormous waste of fish.

I had become use to catching and killing fish for research. I had systematically caught and processed thousands of spotted sand bass for my project and never gave it a second thought. Maybe it was because we specifically targeted the project species using hook and line with great efficiency. We knew where they were and how to catch them. And those species that were caught incidentally could easily be released unharmed. However, when you unleash a web of death into the ocean, targeting a specific species was not possible. Gillnetting is indiscriminate and harsh. It takes a deathly slice out of the ocean community and you cross your fingers that your target species was somewhere buried in the carnage. The useless samples or by-catch are usually counted, measured and tossed overboard dead. And if we were lucky it was just fish. On occasion coastal marine birds and marine mammals would get tangled in the near-invisible net and drown, surrounded by the carcasses of the fish that had lured them there.

A few months after being promoted, I was headed to the boat yard to get the sampling gear ready for another sample set down the coast. I pulled up to the locked gate of the boat yard, unlocked the gate and drove through into the parking lot. I parked near the front entrance of the building and walked down by the dock. The 18 foot workboat was in the water and tied up near the walkway. A bright-red plastic gas can was located near the transom and connected to the engine.

The harbor facility was right on the water and utilized several small docks right next to the main building for tying up small skiffs and storing donated vessels. The storage compound above the harbor held everything imaginable for fixing boats. There were welding benches; stock steel in every length, width and thickness; a full workshop; old outboard motors in varied stages of disassembly; as well as old and new hulls, anchors, old rusty chains and cigarette butts. Everywhere cigarette butts.

Stored at the side of one of the huge outbuildings was every ocean sampling device that could be launched from a vessel. Sediment grabbers, neuston nets, beam trawls, otter trawls, floats and all the assorted gear that comes with these devices were stacked semi-neatly in the shadow of the large tin building. Steel plating, rusted chain, thick mesh netting and colorful buoys made up the pile. If you had no clue on what these instruments were used for, you would have concluded that someone with a medieval dungeon was having a garage sale.

At the edge of all the sampling gear, stacked up close to the harbor, were six large gray bins loaded with monofilament netting. Next to the bins was a pile of rusted anchors and chains, tangled and unfriendly. Next to the anchors, looking like a large bundle of huge blueberries, were a bunch of large diameter buoys, all tied together in some unfamiliar sailor's knots. I reached down and noticed that it had been tied far beyond what was necessary and would only serve to frustrate whoever got to untie the mess. A frustrating joke from Danny, the lead engineer of the facility no doubt. I stepped back and examined the pile of plastic and metal. While the sampling gear was deceptively simple in its individual components, when they were assembled, they formed the very efficient tools of destruction.

I decided to load the lightest part of the gear first. John was slated to help me out on this trip and the heavy gear could wait for him. I kicked the large clump of buoys down to the dock and dropped the whole mess into the front of the Whaler. John showed up a few minutes later and started loading the large bins that contained the nets.

The nets we were using were 150 feet long and 8 feet high and essentially rectangular in shape. Running the entire length of the net on the top was a floating line. A large buoy was attached to each end of this line and helped to keep the net upright and sampling. Running the entire length of the bottom of the net was a lead line, which is a heavy line with a lead core. An anchor and a length of chain are attached to each end of this line. This stretches the net down from the float line and helps hold the net in place. If the entire apparatus is deployed at the right depth, it stretches out like an invisible spider web in the water and catches anything that passes. And I mean anything.

It took about fifteen minutes to load all the gear into the Whaler. As John was moving the anchors to the front, he noticed one of the large buoys piled into the bow. "Whoa, what happened to this one?" He rotated the giant orb so I could see the bottom. The buoy had a large, nasty scrape on it a foot wide and it looked like its ability to hold air would be in question. He continued to poke at it, almost like he was trying to puncture the float.

"Stop poking at it!" I said. We didn't have any extra buoys for the set and if John let the air out of the damaged one, we could only set five nets instead of the required six. He tossed the buoy to the bow and started messing with the anchors. He didn't press me for an explanation, and I happily kept the incident to myself.

On the previous trip I was loading this same gear into the back of my truck for a run to Redondo Beach. The sampling scenario was the same, except that another Whaler awaited our arrival in the Redondo Harbor for a run out to the point to deploy the nets. We were running a little late and decided that to save time, we would attach the buoys to the nets, one on each side, and then stuff the whole mess back into the bins prior to the drive to the harbor. This way we could race out to the point and not have to worry about attaching the buoy lines. To prevent tangling, we always kept the anchors and chain well away from the nets until they were deployed.

Greg and I shoved the over-stuffed bins into the back of my truck and we headed out. All was fine until we hit the highway. I had just merged into the fast lane when something in the rearview mirror caught my eye.

The big blue ball looked very small bouncing behind my truck and its size kept me from making the fairly obvious connection. The reason it was so small was that it had already paid out half the net behind the truck and was merrily bouncing 80 yards behind us. I glanced back at the gear and saw one of the bins vibrating as the remainder of the net was deployed perfectly across four lanes of traffic. The punctuation to the event was the second buoy being launched out of the bin at around 60 mph. Thankfully the trailing traffic expertly avoided the obstacle and after a few horrendous skids and several waving fingers, we were able to gather the whole mess back into the truck with little damage.

John spent a few minutes untying the secret sailor's knot that Danny had left us and then got busy readying the anchors for attachment to the nets. "Hey," he said, not looking up, "toss me the tool." Each anchor and buoy was attached to the net by a turn buckle. After several trips in seawater the small metal clamps were rusted and frozen almost beyond use. The only way to keep them somewhat functional was to spray them liberally with silicone oil and to use the "tool". The tool is a medium standard screw driver, with the last two-inches of the end bent at a right angle for leverage. If you couldn't loosen it with the tool, it was usually welded beyond use and tossed overboard. Without a way to loosen and then tighten the buoy and anchor turn buckles, we were essentially dead in the water. We never left the dock without it.

I grabbed the turnbuckle tool from the center console and handed it to John. I pushed away from the dock and headed for the harbor mouth. We kept two bins up front with John, and I moved the remaining four behind the console and out of the way. By the time we threaded the harbor entrance between the jetties John was hooking up the last net and situating the bow for setting.

Once we left the harbor, I headed south and rolled with the afternoon swell. The overloaded boat and the wind chop kept things interesting and a little wet. While the gear didn't weigh a lot, its bulkiness made negotiating the conditions a little tricky. More than a few times John shot me a dirty look after the bow dipped too low in the trough and sent salty spray over the buoys, bins, anchors and John.

About twenty five minutes from the harbor mouth, we pulled around a long rock jetty and got a little relief from the growing ground swell. Our sample area had been randomly chosen from a group of potential sites, but this area was not new to us. Both John and I had pulled large seines through the surf of Seal Beach a few times before. The length of water we'd be setting the nets in was located between two manmade jetties, approximately one mile apart. The idea was to set two nets each, one parallel and one perpendicular to shore, in 20, 30 and 40 feet of water. When done correctly, a two-man crew could accomplish a set of six nets in about twenty minutes. When done incorrectly, you usually ended up featured on the local news.

John and I studied the surf and swell for a few minutes, but the fading light didn't give us much time to wait for a pattern. I pressed the rubber buttons on the depth finder and the unit blinked on. The screen displayed the flat bottom and a large 45 showed in the upper right indicating the depth in feet. The deep nets would go here, but we'd set those last. To avoid maneuvering around nets during a set, we always started in shallow and worked our way out.

I nudged the throttle forward and we started running in towards the beach. I kept my eye on the depth finder, the growing swell and the shore. Finally, just as the shore break started lifting and dropping the skiff, the screen showed 20 feet and we were ready to set the shallow net. I knew it dropped off quickly here, but I was still a little concerned that we were so close to shore and the thundering waves. We definitely needed to be careful and work fast. The chop had picked up and the four-foot swell was gradually pushing us closer to shore. I was slamming the throttle back and forth just to keep us positioned on the edge.

The idea was simple: as John tossed the buoy off the bow, he would also drop the lead anchor. The Whaler is then backed up slowly and the net feeds out of the bin. At the end of the net, John grabs both float line and lead line and gives them a good tug to stretch the net out. He then drops the trailing buoy and anchor into the water to complete the set.

I made a circle out deep until John had things ready. Once he was set, I moved in towards shore slowly barking out the depth. When we reached the target depth of twenty feet, I told him to toss it. John dropped the anchor overboard with one hand and then tossed the buoy with the other towards the shore. I dropped the skiff into reverse and started backing out. The net started feeding from the gray bin as I maneuvered backwards over the swell. Water sloshed over the transom while John held the bin and made sure the monofilament net didn't get hung up on anything as it fed out. The shallow, perpendicular nets were always the hairiest, because you were backing into the approaching swell and dealing with the shore break.

As we approached the end of the net, I slowed the boat a bit. John grabbed the lead and float line and tugged the net tight. Once tight, he dropped the trailing buoy and anchor over.

The first net had paid out perfectly and before long we had flawlessly deployed five of the six nets. As I maneuvered into position to drop the final net, John noticed a problem. "Hey, we're short one anchor," he said. I dropped the throttle into neutral and started looking around the crowded deck. "Did they all get loaded?" I asked, as I picked up the bins searching for the last anchor.

After a few minutes, John and I decided that neither one of us had bothered to count the anchors to make sure that we had enough. This was a problem. By the time we could get back out here with a replacement anchor, it would be too dark and impossible to set. And we couldn't deploy the net without an anchor on one side. The net would just drift all over the place during the night swinging on the one holding weight. We decided to use the boat anchor. The skiff anchor was a bit larger than we needed, but at this point we had no choice. If we failed to set the last net the data from this station couldn't be used without some serious statistical adjustment. To us, that was unacceptable. A few minutes later, John was attaching the boats only anchor to the last gillnet.

We set the final net and headed out to deeper water to check the set. All the nets looked good except for one. I could tell that the first net we set out wasn't anchored properly and was being pushed by the swell. If we left it, the crew pulling the nets in the morning would likely find it washed up on shore. "We have to pull that net out a little deeper," I said and pointed towards shore.

"No sweat," John responded. We did this all the time, so it really wasn't a problem. But the swell had picked up a bit and I really wasn't looking forward to putting the boat back in near the breakers.

I weaved my way through the set slowly and pointed the boat towards the floundering net. As we approached, John lay down over the bow and grabbed the float line. "Got it!" he yelled as I simultaneously dropped the boat into reverse. He pulled the end of the net up and grabbed the anchor line as well. All we needed to do was have John hold on to the net and slowly drag it out into deeper water as I ran backwards. Once we straightened it out, he could release the net and we could head for home.

I was paying too much attention to too many things. I also misjudged the distance we had to work with between this net and the others we had set earlier. The swell was easing a bit and it looked like we were seconds away from dropping the net and heading back to the dock. That's when the engine died.

I turned around to see a large blue buoy nestled gently next to the outboard. The conditions were quickly becoming dangerous and we were far too close to shore to be without power. The boat's only anchor was safely tied to the last net we had deployed, so we couldn't use it to hold us in position while I untangled the mess. And darkness was minutes away. Despite the drag of the net wrapped around the prop, I could feel the boat move towards shore and the breakers with every swell. The sound of waves crashing against the beach didn't help. If we lose this boat, Larry would never let me out of grad school.

I raced to the back of the boat hoping against hope that only a few turns of the net were wrapped around the prop thus making it easy to free. I lifted the outboard. I could not see the prop. It was completely covered with a tangle of netting, float line and lead line. The thin netting cut deep into the bearings and looked to be there for good. A swell lifted the back of the boat and moved through the stranded craft, testing balance and courage. We had to act and act fast or it was over. "Toss me the tool!" I yelled to John.

John took me at my word and the bent screwdriver came twisting and turning in midair and heading right for my head. I held out both hands and thankfully felt the molded plastic handle land in one palm. I quickly went to work on the tangle. I ripped and tore huge chunks of netting, trying to free the prop. I must have looked fevered and insane from behind as clumps of netting and rope filled the air over my head. John was watching the waves and the approaching shore.

After a few more minutes another swell lifted the back of the boat out of the water, but this time the netting pulled free and dropped back into the ocean. I put the tool into my pocket and leaned over to the center console to trim the engine back down. I hit the starter. The little Mercury came to life like it always did and I carefully dropped it into gear. The froth behind the outboard was a mix of dirty seawater and sand, illustrating that we were in very shallow water and too close to the shore break. I eased over an approaching wave and headed out to deeper water. I could see that the main rope to the buoy had been cut and I know I ripped apart the anchor line freeing the prop. The buoy was still connected to the net by a few strands of monofilament, but the last 12 feet or so of net was completely annihilated.

"That was hilarious," John said from the bow. "You should have seen yourself attacking that net." He laughed.

"Hey, I've seen you swim and I pretty much saved your life."

"Thanks," he said.

We made it back to the dock shortly before dark and tied up the boat. As I got out of the Whaler I noticed an anchor lying on the dock, right where John had left it. "John, you're fired," I said, pointing to the lone anchor.

"Hey, I didn't even know I was getting paid."

"You're not, but you're still fired."

I bought John dinner and a few beers and we laughed about the crazy boat antics like we always did. The following day the retrieval crew gave me a call and said that all had gone well with the net retrieval. They did mention that one of the nets looked like a great white shark had torn through it and ripped it up pretty good.

"Hey," I thought, "that sounded good enough to me." I was just thankful we didn't make the news.

TWO-WEEK NOTICE

About a year and a half after taking the lead on the white sea bass gillnetting survey team, I started to wonder what the next step in my career would be. I did enjoy the contract work, but I felt it was about time for me to move on from the university work and find a permanent job. My body was also starting to feel the effects of the routine netting trips. Constant back aches and cut hands, and the weeks out to sea were starting to take a toll. My body would barely have time to recover from one trip, before the next scheduled trip arrived.

The enormous waste of mass gillnet sampling was also starting to wear on me. I knew somewhere buried in the data sheets was an exact number of fish we had killed during this project, and I knew it was enormous. I had no illusions that my chosen career would ultimately result in the demise of collected specimens in the name of science. But the shear quantity of fish we were killing on each trip was starting to effect how I felt about this particular contract. I guess towards the end I started to wonder if the data we were collecting was worth the loss we were accumulating.

It was the middle of June and we were prepping the *Yellowfin* for another run down the coast to deploy gillnets. With some routine maintenance and the loading of gear, Jimmy, the captain was ready to head south around mid morning.

The large vessel eased out of her slip and headed towards the mouth of the harbor. Danny and I were left on the dock to run the work boats out to meet up with the mother ship out beyond the rock jetties. We each got into a Whaler and started prepping the small boats for the short run to the channel. The two small work boats were to be tied to the back of the *Yellowfin* and towed to the first sample station six hours south in San Diego. For obvious reasons, it is always best to leave the larger vessel unfettered with the small boats as it slowly maneuvers its way out to open water.

As soon as the mother ship cleared the last harbor jetty, it was free from the speed limit constraints of the bay. The roaring sounds of the growling twin diesels spit from the back of the boat as the captain opened them up to put some distance between the boat and the harbor mouth. Danny and I followed suit and raced each other at top speed out to the big gray ship. Danny timed his wake crossing perfectly and vaulted at top speed over the waves left by the big boat. Once across, he turned to watch my attempt, knowing that I didn't have the guts or experience to do the same. As I approached the wake, I dropped the speed of the Whaler and waddled over the waves like a two-year old walking for the first time. Even over the engine noise, I could hear Danny's gravely laugh of victory as he sped off and left me behind. I had lost.

Back aboard the big boat, we headed south down the coast. Our first station on this trip was near San Diego, just outside the harbor. In all the time we've been setting nets to sample the near shore fish species, this station had never gone smoothly. We had lost more gear, time and lunches at this station, than all the other sample areas combined. And while the sample spots were randomly picked within a survey area, it sure seemed like we spent more time in this spot than simple probability should allow. I so despised traveling to this station I had given it the nickname Punta Diablo, Devil's Point.

While we ran, Danny gave the crew, some new and some old, the standard safety lecture. He spent the next ten minutes going through the rules of traveling on the *Yellowfin* and giving us all a quick review of the safety gear and where it was located. He ended the lecture with a stern warning. "Screw up and we'll leave you at the nearest harbor," he said, not totally kidding.

I wandered around the deck checking all the sample gear in the three work boxes, stacked neatly under the sorting table. After dozens of sampling trips it became pretty clear what was needed and what should be in the boxes. After a cursory run through the supplies, I was confident that whatever we needed was on board. I headed to the galley for some breakfast. I grabbed a blueberry muffin the size of a grapefruit and met up with Terri and Jason, two first time volunteers for the trip. I introduced them to the rest of the group and briefly explained to them what would be required during our sampling week.

Since we had a good six hour run to our first station, most of the crew retired to the below deck bunks to grab some rest before the work began. I grabbed my softball muffin and headed out to the back deck. I sat in a lawn chair and enjoyed the solitude of the swaying sea and the early morning sun. The spray from the wake broke up the morning light into a brief rainbow that quickly faded as the droplets fell back into the water. I watched the Whaler bow lines snap in unison as they maneuvered the wide wake of the *Yellowfin*. I was watching the work boats glide in the wake, when one of them decided it wasn't going to take the abuse any longer.

The port side Whaler swayed once more and simply turned outside the wake and reversed course. The current and the progress of the ship made it look like an invisible force had turned the boat around and headed it back towards the harbor. I stood up quickly and whistled for the Captain. Jimmy showed up at the back of the upper deck and peered down at me. I simply pointed to the departing work boat and he went back inside and eased off the power.

After some coordination, we retrieved the independent Whaler and re-installed the bow pin that had come loose. We checked the bow pin in the other Whaler and were once again under way. After that, I retired to my bunk with the rest of the crew.

A heavy swell pitching the boat sideways and muffled yelling from the wheel house had me up and out on deck quickly. When I came out on deck, my first view was of the immense kelp beds of La Jolla off the stern. I glanced towards shore and saw a large wave crash on the distant rocks. A split second later the stern of the boat rose substantial and blocked out the view of the shore as a large swell moved through the vessel. I watched the wave make its way to shore and deposit its energy on the jagged rocks. Conditions couldn't have been worse for setting nets or being in a small boat.

The *Yellowfin* moved slowly at the outside edge of the kelp. The large vessel rose and fell with the moving swell as Jimmy tried desperately to locate a good place to anchor. After several attempts and a few more swells, Jimmy made the obvious choice to anchor further out. The twin diesels fired up and the ship arched out towards the open ocean.

I watched the small skiffs rise over an 8 foot swell and fall in line behind the retreating mother ship. The little white boats looked like toys in the heavy seas. If the big boat can't even handle the swell out here, it's going to be far tougher for the work boats loaded down with gear in close. Just then Danny walked up and asked me who I was sending out to set nets.

The list of potential net setters was shorter than short. In fact it really only had two names on it, besides my own. For safety reasons, the two volunteers were out. I removed myself from the list because I could. Besides, I had set more gillnets than I care to remember for this and other contracts and if I was going to train someone else for this position, I couldn't keep doing it myself. That left just John and Paul. Both were veterans of countless gillnetting trips and both had assisted me in setting nets in the past.

I handed out the work duties and started getting stuff ready. John and Paul both appeared excited at being tasked to set the Punta Diablo nets. Jimmy gave us the tide information and a short time later Danny and I were loading the work boat with all the gear needed for the set. John and Paul were suited up in foul weather gear down at the lower deck eager to run the skiff in to the mouth of the beast. While I held the boat close, they jumped aboard and got the boat ready. In ideal conditions, with trained personnel, a set could be deployed in about 20 to 30 minutes. With the current conditions and John and Paul in charge, we crossed our fingers.

Paul dropped the boat into reverse and pulled away from the *Yellowfin*. John was straddling the gear in the bow and working on the turnbuckles and buoys. Since we were setting at high tide the nets needed to be dropped a little deeper now to avoid serious problems when we picked them up in the morning. Paul nodded that he understood the adjustments and then he slowly backed away from the *Yellowfin*. Once away, Paul dropped the throttle faster than I would've and raced towards the kelp beds.

Just as the sun was setting, John and Paul returned from the set and tied the Whaler up to the back of the boat. They both stated that things had gone smoothly and without a hitch. Since the day's work was done, the crew settled into the galley and enjoyed a hearty dinner. We talked of past trips and laughed way too loudly. I was convinced that if we were moored in a harbor somewhere, we would've been asked to leave. Something I was familiar with on several occasions. But we were miles from shore and no one really cared. After dinner we all turned in early anticipating a full day of work the following day. As we slept, somewhere in the deep, dark sea, monofilament nets were being ripped to shreds over the sharp, mussel-covered rocks and anchors were being twisted and bent into the crevices of the reef.

After a restful night of sleep, a heavy pounding on the cabin door signaled that morning had arrived and it was time to get to work. The

Yellowfin had been rising and falling all night with the swell and the conditions seemed to worsen into the morning. The boat tipped enough for me to catch myself in the bunk room and any thoughts of the conditions improving during the night vanished. I instantly became determined to get our crew out as soon as we could and to pick up the nets quickly. The sooner we could collect the gear, the sooner we could leave.

I woke the rest of the group and told them that we'd be up and out pulling nets in about 20 minutes. I walked through the galley and out onto the back deck. Danny was at the starboard side looking out towards the coast. He turned and gave me a sly grin as I approached. "Check out what your crack team of net setters did," he said, nodding towards the rocky shore. I glanced toward the large kelp gap where John and Paul were told to set the nets the evening before. A large eight foot swell was moving through the gap obscuring my view of the large blue buoys at each end of the nets. As the wave arched and prepared to break, two blue orbs popped out the back of the wave and slid down into the trough. Pure panic gripped me.

"Brace yourself," Danny said, "those are the deep nets."

This was bad. Negotiating the swell and recovering the deep nets was going to be hard enough. Getting in closer to shore in the degrading conditions to pick up the shallower nets was going to be life threatening. And I knew I was going to be manning one of the Whalers. "You ready to go?" Danny said, draping an arm over my shoulder. Ready I was not.

We gathered the group together and explained the predicament. This was without a doubt some very treacherous conditions for retrieving nets and we wanted to make sure everyone was safe. Danny volunteered to go in and get the close nets, and I let him. He also volunteered John and Paul to accompany him, since they were largely responsible for our plight. However, after realizing that this would leave my boat with the two new volunteers, we decided to split them up.

We loaded everyone on the two Whalers and formulated a quick game plan. It was likely that the anchors holding the nets in place would be rocked down pretty good after a full night of tidal surge. It was probably going to take some serious pulling to get them off the bottom. I was hoping that we'd get lucky and be able salvage some of the data from this station after all. I guess I forgot where we were.

Danny, in true sailor fashion, headed straight for the shallow nets. He eased the Whaler over a swell, raced inside and disappeared behind yet another large wave. I motored over to where the deep nets were set and hung outside a bit to see how the swell was moving. I told John and Terri that we would need to do this fast and that the nets may be stuck. John seemed to understand. Terri looked frightened.

After a few minutes, the swell dropped a bit and we headed in to grab the first net. John was on the bow ready to grab the float and anchor, and Terri was standing behind him holding on to one of the bins. I nosed the Whaler up to the closest net and put the boat into neutral. John reached down and grabbed the blue ball and pulled in the net lines. He handed the floating line to Terri and he grabbed the lead line. He took two good tugs on the net and the anchor came off the bottom. John quickly pulled in the anchor and loudly dropped it into the skiff. I kept an eye on the swell as John and Terri pulled us along the net, loading the monofilament, dead fish and all, into the bin as we moved. I glanced towards shore and saw Danny's Whaler way inside the surf struggling with one of the net lines. I couldn't believe how far inside they were.

A large swell moved through our boat and caused all inside to drop what they were doing to catch their balance. John was tugging on the trailing anchor and I could see it was stuck. Just then another swell came through and this time deposited enough water in the boat for me to become concerned. Terri slipped and fell over the bins. John was holding fast to the anchor line and keeping us in place. The surge was spinning us around the net and if we didn't get out of there quick, we'd risk tangling the remainder of the gear in the prop.

"Just cleat it off, John," I said frantically. He wrapped the lead line around the cleat mounted at the front of the boat several times and tied it off. I glanced over my shoulder and saw a fair sized swell coming our way. "Hang on," I yelled. I dropped the throttle and turned the bow into the swell. The engine strained against the anchored line and for a few seconds, we didn't budge. Then the lead line broke loose leaving the anchor on the bottom and we glided over the advancing swell. My heart dropped back down into my chest as we motored out past the hateful surge.

John and Terri stowed the net behind the center console and grabbed another bin for the next net. I looked back in towards shore to locate another blue buoy. I could see Danny's boat struggling with another set of gear further in. I was again amazed at how close to the shore and the breaking surf they were. The Whaler was completely surrounded by white water. All three occupants were pulling on the obviously stuck gear. The boat tipped dangerously as the three worked on the net. Another large wave lifted the front of our boat and raised the vessel a good eight feet before moving through and lightly easing us back into the trough. We watched it travel towards shore and Danny's boat.

I jumped up on the console and whistled loudly to signal the other skiff. The swell was monstrous and completely eclipsed any view of Danny's Whaler. I heard the boat's engine rev up and then sputter a bit. Danny dropped the throttle on the workboat and I could hear the engine roar to life. We all anxiously watched the back of the wave and waited. Just when we thought it was all over, the dark blue belly of Danny's Whaler broke through the back of the wave and headed skyward, moving over the wave and landing hard in the trough. The two occupants in the front of the boat looked shocked and terrified. Danny's steel glare never faded and his cigarette remained lit. That was close.

We spent the next hour retrieving the rest of the gear, carefully. We lost seven of the twelve anchors and two of the nets were damaged beyond repair. The data for the Punta Diablo station had to be scraped because none of the nets yielded complete results. Some were tangled with crab and lobster indicating that they had spent some of their sample time on the bottom. Still others had huge holes ripped in them from rubbing on the jagged rocks for most of the night. In short this station was a huge waste of time and gear, and it had almost killed us.

Back on the *Yellowfin* we dumped all the tangled gear on the lower deck to be repaired and sorted. The gear was a mess. Anchors, mesh, lead lines and floating lines were all balled up and hopelessly tangled. The remains of dead fish were wrapped in every level of the pile. These fish gave their lives for science and now science had to worthlessly discard them. I kicked the pile and felt sick at the waste. I felt like pushing the whole mess overboard and going home. But we still had three more stations up the coast and the nets needed to be repaired and set up for the next stop.

John and Paul started in on cleaning up the gear. Terri soon joined in. I grabbed the data sheets and started cleaning up the wet lab area. Jimmy started the twin diesels and started working the hydraulics to pull the anchor up. I felt relieved that in minutes we'd be on our way and out of here. That didn't happen.

The even tone of the hydraulics became strained and I could tell that we were having trouble raising the anchor. As the pitch became higher, the operator decided to turn it off and investigate. I looked towards the bow and saw Paul leaning over the front of the boat and looking straight down the anchor line.

I walked out to the bow and looked over the side. The anchor chain dropped straight down from the bow and disappeared into a huge cloud of brown kelp. The hydraulics couldn't lift the 1,000 pounds of slimy plant and the anchor out of the water together. After a brief discussion, Paul appeared at the front of the boat with a large carving knife. He jumped up on the bow and carefully slid down the anchor chain to the anchor. Most of the algae was underwater, but a few dozen thick stipes were wrapped securely around the top of the anchor and within easy reach. As Paul bent down to start cutting, Danny told him to be careful. Paul took exactly three swipes at the plant before he laid the blade across the back of his hand. The cut was deep enough and long enough for him to abandon his cutting tasks.

With some help and a few paper towels, we lifted the battered kelp warrior back aboard and re-grouped. Danny and Jimmy decided that maybe if they adjusted the hydraulics below deck, they could lift the anchor out of the water and the vile weed could slide back into the depths.

Danny went below deck to do some adjusting. After a few minutes, the hydraulics came back on and the anchor chain tightened. The thick chain links began to slowly march on board. We all took a few steps back when the whine of the winch strained and got a lot higher. At the point where the anchor began to leave the water, the hydraulic whining stopped abruptly and the heavy blob of algae and anchor slid back into the water with a substantial splash. Jimmy appeared at the top deck with a stunned look. "What the hell happened?" None of us knew. That is until Danny came up on deck, completely covered with dark hydraulic fluid.

Just when the hydraulics was dealing with the load at its heaviest, a hose let loose and sprayed Danny with the fluid. He wasn't injured, but this just added to the already mounting incidents on this trip.

After several more attempts with an improvised tool, which was essentially the same knife Paul had used to cut himself, duck taped to a broom stick, we were able cut loose the anchor. After Danny cleaned up, he repaired the blown hose and in no time the anchor was once again resting in its cradle at the bow. There was no sweeter sound than that of the twin diesels firing up and motoring us the hell out of there. I looked out at the undulating kelp bed that had so often abused us. I was glad to be leaving. I had no way of knowing that I had just set my last gillnet at Punta Diablo.

John and Paul, with some help from Terri, the volunteer, had made good time cleaning up and repairing the nets. As I watched them pulling dead fish from the pile, I had to wonder where the other volunteer was. I took a step out on deck and looked around. I looked to the upper deck and saw a pair of bare feet slightly dangling over the edge of one of the lifejacket storage bins. While John and the group had been slaving on the huge gear ball, I had been gathering all the equipment needed to rebuild three of the nets and searching the holds for anything we could use as spare anchors. Danny was busy rigging the good nets with breakaway lead lines so that we wouldn't rip the bottom of the nets out when the anchors got stuck. The cook was getting lunch ready and Jimmy was driving the damn boat. In short, everyone was working except Jason.

I walked up to the wheel house and took a look outside. Jason was laying out on one of the storage bins catching some sun. "How long has he been out there?" I asked. "Ever since we started messing with the anchor," Jimmy said. During our pre voyage meeting back on campus, I told both Terri and Jason what was expected of them on this trip. If there was work to be done, they needed to be doing it.

I walked out and sat down on the second storage bin next to the one he was laying on. Jason heard me approach and looked up. I decided to give him one chance. I had already made it very clear to both volunteers that this was no pleasure cruise. With the disastrous first station taking its toll on our gear, we'd have work to do all the way to the next station.

I politely restated the boat's work policy and Jason, somewhat reluctantly, lifted himself off the bin and made his way to the back deck to help out. I watch as Jason inserted himself into the net crew and did the absolute minimum to lend a hand. I had no doubt that before this cruise was over, I'd be discussing work ethics again with Jason.

Lunch was served as we ran north. The trip was smooth and sunny and the mood had definitely lifted since we had left La Jolla. The big gray ship eased into the approaching swell and settled into a calming cadence. She pushed the grumpy sea aside and rocked us in a protective cradle. I had lost count of how many trips I had made on the *Yellowfin* over the years. She was s solid ship, and even on some of the rougher trips, I had never questioned her sturdiness.

Our next station was still more than two hours away and it looked like the nets would be ready to set by the time we got there. The repair job had been intense, but with all competent hands on deck assisting, John and Paul were sewing up the last panel of the last net as we ran north. Jason had once again disappeared during the real work and I didn't have the time or the patience to force him back to work. We did have spare sample gear on board, but most of that had been parted out to make one complete set thanks to the forces at Devil's Point. With the main gear made up of parts and pieces of the spare gear, there really was no more room for error on this trip.

We arrived at our next station about four hours before we needed to set the gear. We spent the afternoon double checking everything. I helped Danny finish up the breakaway rigs for the lead lines. This would save us from major net repair if things went awry again. To avoid a repeat of the last station, I decided to deploy this set myself. John felt bad about the previous station and volunteered to help deploy again. I assured John that with the wacky tides and the rough surf, it would have been a tough set for anyone.

About an hour before sundown, John and I started loading the six gray net bins into the Whaler. Each bin had the two large buoys sitting on top of the gillnet already attached to the gear. The anchors were placed side by side in the bottom of the boat, the rusted links and turnbuckles coiled on top for easy access. To avoid tangles with the angry metal gear, the anchors were kept separate from the bins and attached to the nets at the last minute.

Danny held the bow line, waiting for John and me to organize the gear. "You ready?" he asked.

"Not quite," I said. I looked up and saw Jason once again sitting on the life jacket bin, enjoying the sunset. During the gear prep, he had once again disappeared and was essentially useless. I felt beyond annoyed. It was time to put him back to work, or at the very least interrupt his free time.

Jason didn't seem too happy about piling in on the gear-crowded boat. I didn't care. We were already two days into the trip and he had gotten off way too easy work wise. John could easily deploy the nets himself and he was good at it, but I wanted Jason to at least help out. As we headed for the station, I somehow knew he'd be little more than ballast for this set.

The gear had deployed perfectly, no thanks to Jason. John had done a great job and seemed to work around the useless volunteer. After we were done, Jason couldn't get off the skiff fast enough as we came up beside the *Yellowfin*. Instead of jumping off with the bow line and tying us off, Jason jumped aboard and made his way towards the galley and dinner. "Hey," John yelled, as Jason reached the upper deck. The heavy bow line almost hit him in the face when John tossed it on deck. Jason grabbed it and tied us off quickly, then proceeded to dinner. "Idiot," John mumbled. "Where'd you get this guy?"

"Don't worry," I said, "he'll be out there pulling on dead fish early tomorrow."

That night I climbed into the dark bunk with an aching back. After several minutes I finally found a position where I felt comfortable enough to sleep. I flexed my sore, peeling hands as the big ship moved with the sea. I thought about all the time I had spent on the ocean in the last five years. I had missed my share of family events due to the obligations of marine science. At one point I had spent three birthdays in a row on board this ship. And I had to admit, the excitement of going to sea just wasn't there for me anymore. The work was backbreaking and at times dangerous. Recently I had become sick of the incidental waste. Hundreds of pounds of dead fish were frequently just counted and tossed back into the sea. I started wondering where the line was between studying a marine animal to better manage the species, and sampling it towards the edge of oblivion. With more experience came more responsibility. Now that I was the lead, I was also going to have to continue to scrape up volunteers for future trips, and deal with them when they invariably turned into worthless punks out to sea. Jason wasn't the first deckhand who had turned into a waste of space as soon as we left the dock. And he certainly wouldn't be the last. I stretched as far as I could in the confines of the cramped bunk and felt light headed. I felt like the job was beating me up.

The morning arrived on a flat, calm sea. With three workers per skiff, we collected all six nets in less than 40 minutes. The fully loaded bins were taken back to the *Yellowfin* and sorted. The nets were examined for holes and repaired if needed. While I attended to the data, John and the volunteers rinsed off the nets, loaded them back into the bins and set them on the lower deck for the next station. A smoother set I could not remember.

We had all the gear aboard and the nets already stowed for the next set by 10:00 AM. The work skiffs were tied up to the A-frame and harmlessly drifting behind the mother ship. Once again the sweet sound of the diesels signaled that lunch would be on the move and if you had to judge the success of the trip by this last station, you could say that things were going well. All the earlier events of the trip were just a bad memory.

The run to the next station was calm. We had sandwiches on the back deck for lunch and we all enjoyed the halfway point of the trip. Jason was once again sacked out on the upper deck. He had help pull in the nets earlier, but that had been about it. Danny had made it clear that he didn't want him on his work boat and if I had a choice, or one more person to assist, I wouldn't want him on my skiff either.

We arrived at the next station a few hours early. The crew had the gear cleaned and repaired and ready to set hours before we arrived. Terri, the other volunteer, had performed well with the experienced crew, while battling bouts of sea sickness. His willingness to continue to assist after heaving his lunch over the side was admirable. Jason, on the other hand, was completely worthless. And most of the crew, including Jimmy and Danny, made very sure Jason was aware of his status, especially at meal time.

In the enclosed community of a vessel, once out to sea, the entire ship must work as a unit. If someone decides to not play by the rules, the crew can turn on that person rather quickly. Volunteers on these survey cruises don't get paid and essentially worked for their meals. Most onboard felt like Jason didn't deserve to participate in any galley activities. When meal time came, his portions were scrutinized and he was verbally battered if he decided to actually eat in the galley. His non-existent work ethic had stirred the boats regulars into a bloody froth, and he was either ignored or given the verbal equivalent of a keel-hauling on a daily basis. As the voyage progressed, most worked around Jason or completely ignored him.

After each station was complete, I had to fill out the data sheets to summarize the catch. The main focus of the contract was to catch white sea bass and assess if they were hatchery fish or not. The data for the San Diego station couldn't be used since half the nets were destroyed. Despite the waste, we tallied the species we had collected in the working nets and no white sea bass had been caught. Today's station resulted in more of the same. Hundreds of dead, non target species were collected and no white sea bass had been sampled.

On past trips it wasn't unusual to go the entire trip without seeing a white sea bass in the nets. The ocean is a big place and trying to collect a handful of fish that had grown up inside a hatchery facility was a tall order. When we did collect the target species, we'd all excitedly gather around as the magic wand that detected the small wire tag inside the fish's head was waved over the sample. And on trips when we were lucky enough to collect a re-capture, we were on top of the fish sampling world.

Since we had arrived on station early, we spent a few hours fishing the nearby kelp beds until it was time to set. We hooked up regularly and for a short time the rigors and pain of the gillnetting work was completely forgotten. Most of the fish were barely legal and as soon as they were unhooked, all the caught fish were released.

On my last cast of the day, I hooked and landed a large opal eye. The olive-green fish danced around enthusiastically in the sorting table while I tried to pry the lure from its face. For the most part this species feeds on algae, making it an unlikely catch on any artificial lure that imitates a living creature. They were also extremely sturdy, often being the last live fish in the net when they were unlucky enough to be caught.

I pried the green lure from its mouth as it erected its spiny dorsal fin. It flailed again, driving one of the spines into my index finger drawing a dot of blood. I didn't even flinch. My hands had become tools, the scrapes and scratches the sign of instruments used often. The infected pain would come later. I gently picked up the stressed fish and tossed it overboard. It hit the water with a slap and was gone. I looked over the side. "Stay away from the nets!" I said, as I wiped the blood on my jeans.

The evening set went well and Jason was not happy that he had to tag along with me and John. I didn't care. I had decided that I was going to work him good and hard while we were out here. And if that meant putting him on any job for the next few days just because, that's what I was going to do.

About ten minutes after the crew got up, I was at the helm of one of the work boats with an apple in my mouth warming up the outboard. Jason and John were both gorging themselves on huge, softball-sized muffins at the bow.

Danny's boat peeled away from the stern of the *Yellowfin* and headed in to retrieve the shallow nets. John pushed our skiff off and I motored around the front of the mother ship to pick up the deep nets. The sea was flat calm and the morning sun was already warming us. The conditions were perfect and hauling in the nets would be nowhere near as dangerous as the first day. I nosed the Whaler up to the first buoy. John reached out and pulled the large float into the boat. Jason slid the bin up to the bow and they both began pulling the net in, feeding it into the container as they hauled. I kept the boat barely in gear and guided us along the net. We hadn't gone ten feet when it became very clear there was a problem.

Both John and Jason were having trouble pulling the net aboard. John was giving it his all and Jason was just giving in. "It feels like it's loaded," John said, straining against the float line. I looked over the side and could see the net disappear into the deep green water. I could also see several light green forms caught in the mesh. The closer ones had the distinguishing spots on the dorsal area of the opal eye. And all within view were still alive.

The three of us pulled on that net for 30 minutes. We gradually fed the loaded mesh into the front of the boat. We strained against the tremendous load, trying hard to maintain balance and not sink. As expected, the net was almost completely plugged with hundreds of near dead and dying opal eye, the heartiest fish of the kelp bed.

We took a quick break to re-adjust the load and give our backs a rest. I stepped over the writhing mass in the front of the boat just to get away. I stretched out my aching back against the center console. As I bent down, I noticed a large opal eye lying in the bottom of the boat, free from the net. I picked up the still live fish and put it back in the water. As I stood, I saw Danny's boat cruising by about 40 yards away, all three of his nets neatly tucked in the bow. Danny's voice quickly came over the radio. "We're all done. How're you guys doing?" he added in a far too cheery voice, his raspy cackle trailing off as he headed back to the *Yellowfin*.

I looked up towards the bow. John was once again tugging on the overloaded net. Jason was stretched out in the bottom of the boat, his eyes closed. His feet were resting comfortably on the big ball of net and fish in the front of the skiff. The stench of dead fish filled the Whaler. "Ah crap!" John spouted. I looked to the bow. "The lead line just broke," he said.

We spent another hour in the small boat gathering the rest of the gear. By the time we got back to the *Yellowfin*, Danny's crew had already worked up the first three nets and was just cleaning up. My crew decided to work up our nets in the Whaler at the side of the mother ship. That way we could return any live fish we encountered back to the ocean quickly after we collected the data. I tied up to the rail and noticed Jason grab the clipboard with the data sheets off the center console. In the world of field science it doesn't take anyone long to realize that data recording can and often is the easiest job during the work-up. I ripped the board from his hands without saying a word. He just stood there. I took a step closer to him and pointed to the net choked with dead and dying fish. As Jason reluctantly started to untangle fish, I could see John behind him smiling.

Over an hour was spent untangling the fish from the gear. Jason had to be warned more than once that ripping the fish out of the mesh and damaging the net was not acceptable. The last net was hopelessly tangled with dozens of sheep crabs letting us know that the net had spent some time on the ocean floor during the set. Their long spiny legs were wrapped with the mesh to form a kind of thorny puzzle. I lost count of how many times my sore hands were jabbed with the sharp crab legs during the work-up.

Just before lunch, we finished processing all the fish and cleaning up all the gear. The lead line on the first net was repaired easily and all the others appeared to be in good working shape for the next set. We lined up the bins and once again loaded the nets back into their containers in preparation for the last station of the trip. I leafed through the data sheets and noticed that we had collected 205 opal eye in the first net. I knew we were able to return almost half of that back into the water alive, but I still hated the waste. Jimmy started the vessel and pointed her north for our last set. We eased by Newport Beach and soon the familiar drone of the engines meshed with the slow rocking of the ship. We were almost home.

During the run north, I took the opportunity to once again remind Jason why he was here. He had made enough comments during the work-up this morning for me to become irritated. I didn't say anything then, but John had been blunt enough a few times to shut Jason down. After his second warning, I did my best to avoid Jason for the rest of the trip.

That evening John and I set without Jason and all went well. We ended the day fishing for sharks off the back deck and feeling glad that early the next day we'd be home. The quiet evenings always gave me time to reflect. This trip had been hell. And even though I had experienced other trips with problems, this one seemed to go anything but smoothly.

John and I enjoyed the solitude of the ocean. John was a hard worker and a good friend. I always enjoyed having him along on the sample trips. He knew what he was doing and also knew how to have a good time.

The lights of the shore beamed to us like little diamonds. I felt a slight wave of envy for those that got to sleep in their own beds and move around on dry land. My time in the marine program and out to sea was valuable to me. I had learned more than I could imagine and developed relationships that would last a lifetime. On this trip we had run into problems, but the issues were no different than what I've seen on other trips. The frustration I was feeling wasn't because of the trip or the contract, it was centered on me. I wasn't being trained anymore or learning new things. I was by my own admission the lead net monkey and this was most certainly not where I wanted to end up. The job had become routine and had always been back breaking. The carnage accumulated in the name of science had been an irritating and disgusting thorn in my side and now that thorn had become infected. Inside myself I knew that I would never be ok with the useless death of what we collected. My once excited and infectious attitude had degraded into a surly mope that wanted nothing more than to get home almost as soon as I stepped out on the deck. It wasn't fair to the people that chose me to lead the sampling, the people I worked with, and more importantly, it wasn't fair to me. That evening on the back deck of the *Yellowfin*, fishing with my friend John, I knew that I was done working on this contract.

Early the next morning John and I were in the Whaler once again slowly drifting towards the first buoy for the last time. I had decided to leave Jason on the big boat to clean the stainless steel wet lab where we sorted our samples. Short of drying the hull of the *Yellowfin* with paper towels, this was the best made-up job I could think of. Jimmy volunteered to make sure Jason stayed busy while we were all out working.

John reached down and grabbed the big blue float. He pulled in the float line and anchor line and dropped the anchor onto the deck and started pulling. Almost instantly he was met with solid resistance. I grabbed the line and together we tried to pull the net in. Slowly but surely it started coming in and the reason for the resistance became clear. The net was completely plugged with heavy sticks and branches that had washed down from a nearby creek. There was no way we were going to pull the loaded net aboard as is. We would have to clear it as we went.

We spent hours clearing the nets and finding all sorts of odds and ends caught in the mesh. We found shoes, a purse, two shirts, a pair of pants and a sock tangled in the monofilament. As the clothing in the bottom of the boat started to pile up, John and I both started to wonder if we'd be pulling a body out of the net soon.

At the end of the last net, a large ball of kelp had settled between the netting and the buoy. I gave the net a few good tugs and the big brown ball floated loose. Just before it popped free, a pink plastic Easter egg floated out from under the kelp and bobbed to the surface a few feet from the boat. I reached into the water and grabbed the egg and cracked it open. Inside was a soggy ten dollar bill. "Hey," I said to John, holding up the money, "I got a raise!"

Since the netting had been plugged with debris, not many fish had been collected. The data collection only took a few minutes and the last summary sheet was filled out. Out of all the nets we set on that trip, we did not collect a single white sea bass. I stowed the data sheets and helped clean up the gear. We shoved the bins below deck, stacked the anchors as neat as they could be stacked and closed the hatch door. The trip was over.

We approached the harbor mouth, and Danny and I ran the Whalers in back to the dock. We tied them up as Jimmy backed the *Yellowfin* into her slip. I made my way back to the boat to offload my gear and head home before traffic trapped me. I thanked the crew and the cook. Danny and I briefly discussed how we were both pretty sure Jason would never be aboard the *Yellowfin* again. He had been the first to depart once the boat was tied up and as far as I knew he was already gone. If I had known he was waiting for me in the parking lot, I probably would have made him wait a lot longer.

I walked up the aluminum dock walkway to the parking lot carrying my gear. My lower back was screaming and I adjusted my pack a few times to ease the pain. I threw all my gear in the bed of my truck and reached into the side pocket for my keys. As I turned around I saw Terri and Jason approaching the truck. They definitely looked like they had something on their minds. Well, this ought to be good.

Terri had done a good job on the boat; Jason had not. If in future trips Terri showed an interest to return, he was welcome and I had made that clear to him before he had departed.

Terri stayed a few steps back and Jason did all the talking. He started by saying they had both worked hard on the trip and had given up five days to assist. I nodded slightly. He went on to state that the work was more physical than either had imagined. I glanced over to Terri, who instantly averted his glance.

"So, we were wondering if we could get paid a little something for the effort," he said.

I couldn't believe it. I again glanced to Terri, and he looked embarrassed. "Whose idea was this?" I asked, directing the question to Jason.

"Well, we –"

I cut him off before he got started. I had been keeping it in for the last four days and I had had enough. I don't really remember what I said after that. I only know that I let loose with a slew of obscenities in a complete rage. I took two steps towards Jason, putting my face less than an inch from his. I directed all my anger and disappointment at the slacker, although I did see Terri take a step back. I caught the reflection of my unshaven face and wild boat hair in Jason's sun glasses, mouth open and scowling in hateful rage. I scared myself.

When things calmed, I again stated that volunteers don't get paid. "And YOU" I said, pointing directly at Jason's face, "are lucky we didn't leave you someplace down the coast."

Both stood there for a second and then turned to walk away. I stood there shaking my head. I got into my truck and headed home. What an idiot!

Back home I tossed my gear in the corner and collapsed on the bed. I lay there staring at the ceiling, still swaying from the movement of the boat. I thought about the trip and all that had happened. I looked at my swollen hands and adjusted my sore back. The altercation in the parking lot with Jason was just the icing on the cake. I knew I was done. I reached over and grabbed the phone and called the contractor and gave him my two-week notice. Since the trips were monthly, a two-week notice was pretty pointless, but he understood. My gillnetting days for this contract were over.

After I hung up the phone, I lay there on the bed with my stomach growling. I reached into my pocket and pulled out the crumpled and now dry ten dollar bill I had found in the plastic egg. "A little severance pay," I thought. As I held it up to the light, I wondered how it had found its way to the ocean, almost two miles offshore. "Well," I said, "I don't know where you came from, but I know where you're going." I reached over to the phone again and this time I ordered a pizza.

GETTING OUT

Before I even considered graduate school, I had a vague understanding of what it took to get accepted into the program. There was enough idle chatter in the fisheries lab from Larry and the other graduate students for you to get a general idea. From what I gathered, the most important step was finding a professor who would actually accept you into their program. Kind of like being invited to a private, two year long party. Since their spots were limited to funding and university space, you usually had to wait until current students moved on. A few letters of recommendation were also considered gold stars for potential applicants. Convincing other professors to put in a good word for you went a long way in convincing the university that they were making the right decision in accepting you. Good grades in classes with similar project discipline were highly recommended, but not completely required. If a professor wanted you bad enough, he or she could petition to have certain questionable marks accepted. Or they could simply accept you and then make you re-take the classes you had trouble with. This generosity is tempered with the cold fact that now in graduate school, there were only two grades that were acceptable, and anything lower could get you ejected from the program.

The last step was coming up with a good graduate project. Usually, if you came to the table with an interest in the program field, your advisor could help you design a project that not only fit your interests, but fulfilled the requirements of the university and your professor. For these reasons, the true specifics of your project were usually ironed out after you actually got into the graduate program. Most prepared students came in with a few working ideas that fit the area of interest of the professor. It was not a good idea to come in and simply state, "So, what do you want me to do?"

For a new graduate student, the hoops of entering the world of the advanced degree are well understood and easily accomplished if you met all the requirements. Once you're in you feel like the most popular guy at the party. And while nothing much has really changed, you do feel different. You have been plucked from the immense sea of undergraduates and set on the next level to shine. You definitely feel special. Unfortunately, what's lost in the festivities of becoming a new graduate student is the mountain of work that is required to get out.

Once you cross over, your first order of business is to identify a research committee. It was a given that Larry, my graduate advisor, would occupy one of the three positions. The committee is made up of professors who are either familiar with you or your potential work. They read your thesis drafts, provide you with comments or edits, and hopefully sign off on your project when you finish. The committee sat like sentinel wise men to your future, deciding if you have completed the required work for your degree. They could literally let you pass to harvest the employment fruits of your hard work, or send you back to the lab in sour defeat. Since, you as a student got to choose your committee it didn't take a genius to figure out that just maybe you wanted some friendly faces on the other side of the table when they were deciding your fate.

The only stipulation for the three important seats was that the individuals needed to be professors in disciplines close to the student's discipline. I undoubtedly pushed that criteria to near its breaking point when I petitioned to have my under graduate geology professor sit on the committee for a marine biology project. However, after filling out the justification for the seat appointment, drawing direct correlations between plate tectonics and sea formation, I had a rock man sitting right beside Larry.

The third seat was begrudgingly accepted by a marine ecology professor who specialized in the study of algae. Too immature to understand the importance of seaweed, a label that will send violent shudders up the spine of true phycologists, I found my algae class boring and pointless. During a laboratory practicum, having prepared for it during the halftime of a Los Angeles Laker's game the day before, I peered into a slide of a star-shaped algae biospore. Awash in the rapture of their overtime win the evening before, the only question I was prepared to answer regarding the pink structure, is what color it was. I was fairly certain my answer of Algae Death Star, surrounded by penciled laser beams shooting from the words was nowhere close to correct. Although, writing this now, I would have to imagine that Professor Bob would be pretty impressed that I still remember what an algae biospore is. Hell, I'm impressed. And so with some convincing from Larry, Bob occupied the third and final seat of my research committee, and with seaweed, rocks and fish represented, I figured I had all bases covered.

A few weeks after assembling the committee, I was unexpectedly presented with a two-year binding contract in a seedy little hotel room in Monterey, California. Our new graduate group had traveled with Larry to drag beam trawl nets around the coast for a small fisheries contract. It was a two day trip and after a full day of pulling on gear, Larry presented each of us with the daunting document. I personally felt trapped. The contract essentially outlined what was expected of new graduate students by Larry and the university. And while I'm almost certain none us thought we'd be officially entering graduate school by signing a contract in a small room with free HBO and vibrating beds, one by one, we obediently signed the document. For some reason, afterwards, I felt dirty.

Buried deep in the language of the contract, was a requirement that went undetected by me during the signing. I was aware that a thesis defense sat like a huge gray cloud at the end of my graduate work, but an additional presentation at a major fisheries conference was something I wasn't aware of. After the first year, I'd hear about some conference in Canada, but I always figured it didn't concern me. I was wrong. In the spring of 1994 I found myself on stage and at the podium in front of five hundred fisheries scientists and graduate students. I can remember feeling very nervous as I ascended the stairs to speak. I then felt an overwhelming confidence come over me as I reached the podium. I looked out over the crowd for about three seconds before saying anything. I knew that the next fifteen minutes of their lives belonged to me. I smiled and began my presentation. Of the eighty-eight talks in my fisheries category, my presentation was voted in the top three.

While it is very easy to get a romantic view of being a graduate student in and amongst undergraduates, this lofty appointment can become tarnished if you take too long to finish. Depending on the research and the availability of samples, some graduate students can become so enamored with their graduate moniker that they freely extend the two year degree into seven. I'm sure lazier quasi-adults would further stretch the boundaries of time if most universities didn't put a seven year time limit on acquiring a Master's degree. Having known some of the long timers, and having been within earshot of their advisors when they weren't around, it appears that the graduate student luster begins to dull in the eyes of all whom once saw you as promising and bright at around year five. I assumed that the last two years given you towards the deadline were simply added so you'd have time to get your shit together, clean up your area and get out.

There was one looming task that awaited every graduate student like a huge, fist-sized helping of Brussels sprouts on a kid's plate. That's your thesis defense. No matter how long you stretched your project, your defense sat there like a bully at the end of the block waiting for you. You could have the best project known to man, discovered results that would set the science world on its ear, but all that could be derailed and set ablaze in a fire of wasted time if you didn't impress during your thesis defense. It was the great equalizer. In the scads of science disciplines and the equally diverse project subjects, the one thing that tied all graduate students together, was their thesis defense.

I have been present during some superior, hour-long dissertations that concluded with such a roar of applause over the event that if I wasn't already in graduate school, I'd run out and enlist. I've also observed a handful that made you wish a meteorite would crash through the roof of the auditorium, taking out just the speaker. I've also been a part of a clandestine impromptu operation that fed a crucial piece of research data to the presenter, saving his talk from academic catastrophe.

I was sitting in the back during John's defense when someone asked him how old the spotted sand bass was when it first reached maturity. In the world of fish science, if you get up and profess to be the expert on a species, you better damn well know everything about it. Age and size at first maturity are basic staples in fisheries science, and are the field's knowledge equivalent of someone asking you your shoe size. If at the precipice of your graduate career you stumble or fail to answer one of the easiest questions about your fish, your once promising graduate career can vanish in a wisp of cartoon smoke.

The stunned silence from the speaker shut down the proceedings. The only sound growing throughout the room was that of an imaginary space rock hurdling through the atmosphere, headed for the stage. John is one of those rare individuals that is so confident in his intellect that he doesn't feel the need to fill any verbal silence while formulating an answer. He simply stares at you until the answer comes pouring from his mouth like a boiled-over pot. However, when the age at maturity question came, John's pot began to burn. He stared blankly at the audience, and then looked up towards the ceiling. John was in trouble. I held my right hand over my chest and spread out all five fingers, clearly indicating the number five. He glanced at me and in a confident voice stated that spotted sand bass reach maturity at age five, leaving the units off the answer. I applauded his ambiguity. Or his complete cover of not knowing what the units were.

Expanding on my Canada presentation and polishing up the end results of my study, I proudly presented my thesis to a packed auditorium in the summer of 1995. I don't remember much of what was said or the questions that I was asked, but I do know that I got through it. Once I finished, the three committee members met separately to discuss my fate. I waited in the hallway outside with family and friends. Even though I knew that I had done absolutely everything in my power to answer my hypothesis question, I felt the most nervous about it while waiting in that hallway. About twenty minutes later, my hand-picked committee came out carrying champagne and glasses. I was done.

Looking back, I still marvel at my perseverance and tenacity while in graduate school. I put in long hours and longer days to get all the work done and to meet the stringent requirements of both my professor and the university. I can say that if I had to do it again, I probably couldn't. At that time, I was more focused, more thirsty and willing to do whatever it took to finish the project. I've noticed that as I get older, some of these traits seem to dull a bit. And things that I thought were important back then are just not as important to me now. I guess that's why we go to college in our twenties and not in our forties.

In the fifteen years since my time at Northridge, I have worked as a marine biologist for a marine consultant, a fisheries manager for a non-profit marine enhancement project and as a fisheries biologist for the California Department of Fish and Game. The extensive training and work experience I received under Dr. Larry Allen set me far apart from anyone else applying for these positions.

As I moved through life and gathered more experience, I was amazed at how diverse this field can be. I once thought that when I left graduate school, the adventurous part of my training would be over. That has been far from the truth. I have continued to live adventures that clearly illustrate that I have chosen a fantastic career, and I am confident it is where I belong. And what I can state emphatically, is that the story by no means ends here.

Made in the USA
San Bernardino, CA
02 April 2017